Lecture Notes in Mathematics

C.I.M.E. Foundation Subseries

Volume 2272

More information about this subseries at http://www.springer.com/series/3114

Fondazione C.I.M.E., Firenze

C.I.M.E. stands for *Centro Internazionale Matematico Estivo*, that is, International Mathematical Summer Centre. Conceived in the early fifties, it was born in 1954 in Florence, Italy, and welcomed by the world mathematical community: it continues successfully, year for year, to this day.

Many mathematicians from all over the world have been involved in a way or another in C.I.M.E.'s activities over the years. The main purpose and mode of functioning of the Centre may be summarised as follows: every year, during the summer, sessions on different themes from pure and applied mathematics are offered by application to mathematicians from all countries. A Session is generally based on three or four main courses given by specialists of international renown, plus a certain number of seminars, and is held in an attractive rural location in Italy.

The aim of a C.I.M.E. session is to bring to the attention of younger researchers the origins, development, and perspectives of some very active branch of mathematical research. The topics of the courses are generally of international resonance. The full immersion atmosphere of the courses and the daily exchange among participants are thus an initiation to international collaboration in mathematical research.

C.I.M.E. Director (2002 – 2014)
Pietro Zecca
Dipartimento di Energetica "S. Stecco"
Università di Firenze
Via S. Marta, 3
50139 Florence
Italy
e-mail: zecca@unifi.it

C.I.M.E. Director (2015 –)
Elvira Mascolo
Dipartimento di Matematica "U. Dini"
Università di Firenze
viale G.B. Morgagni 67/A
50134 Florence
Italy
e-mail: mascolo@math.unifi.it

C.I.M.E. Secretary
Paolo Salani
Dipartimento di Matematica "U. Dini"
Università di Firenze
viale G.B. Morgagni 67/A
50134 Florence
Italy
e-mail: salani@math.unifi.it

CIME activity is carried out with the collaboration and financial support of INdAM (Istituto Nazionale di Alta Matematica)

For more information see CIME's homepage: **http://www.cime.unifi.it**

Tristan Buckmaster • Sunčica Čanić •
Peter Constantin • Alexander A. Kiselev

Progress in Mathematical Fluid Dynamics

Cetraro, Italy 2019

Luigi C. Berselli • Michael Růžička
Editors

 Springer

Authors
Tristan Buckmaster
Department of Mathematics
Princeton University
Princeton, NJ, USA

Peter Constantin
Department of Mathematics
Princeton University
Princeton, NJ, USA

Sunčica Čanić
Department of Mathematics
University of California
Berkeley, CA, USA

Alexander A. Kiselev
Department of Mathematics
Duke University
Durham, NC, USA

Editors
Luigi C. Berselli
Department of Mathematics
University of Pisa
Pisa, Italy

Michael Růžička
Abteilung für Angewandte Mathematik
Albert-Ludwigs-Universität Freiburg
Freiburg, Germany

ISSN 0075-8434   ISSN 1617-9692   (electronic)
Lecture Notes in Mathematics
C.I.M.E. Foundation Subseries
ISBN 978-3-030-54898-8   ISBN 978-3-030-54899-5   (eBook)
https://doi.org/10.1007/978-3-030-54899-5

Mathematics Subject Classification: 76-06, 35Q30, 76A02, 76F02, 74F10

This Springer imprint is published by the registered company Springer Nature Switzerland AG.
The registered company address is: Gewerbestrasse 11, 6330 Cham, Switzerland

Preface

Fluid mechanics is an extremely active area of science, with interdisciplinary interplay of pure and applied mathematics, physics, geophysics, engineering, medicine, and biology. The lectures delivered at the CIME School on "Progress in Mathematical Fluid Dynamics" in Cetraro, June 2019 were devoted to some foundational issues regarding the Euler and Navier–Stokes equations and related reduced models (as existence, uniqueness, formation of singularities, and vanishing viscosity limits), but also to the problem of a fluid interacting with moving domain with elastic boundaries. More precisely, the questions addressed span from the basic problems of existence of weak and more regular solutions, the analysis of potential singularities, qualitative and quantitative results about the behavior in special singular scenarios, to the modeling and numerical analysis of the solutions of FSI problems. Each of the four speakers gave a set of 6 h of classes, with introductory material and more specialized content in the final lectures. As an additional activity, a session of short communications has been organized during the school, where some selected senior students presented their own ongoing research results.

Here a brief summary of the contributions in this volume:

The volume contains the lectures by Tristan Buckmaster (Princeton University), Sunčica Čanić (University of California, Berkeley), Peter Constantin (Princeton University), and Alexander A. Kiselev (Duke University).

Prof. T. Buckmaster in the contribution: "Heuristic Approach to Convex Integration for the Euler Equations" presents several recent groundbreaking results concerning the use of convex integration to construct wild weak solutions of the incompressible Euler and Navier–Stokes equations. This chapter contains a quite informal introduction focusing on the main ideas, more than on the technicalities, and will be particularly useful for students starting their work in this field.

Prof. S. Čanić in the contribution: "Fluid–Structure Interaction with Incompressible Fluids" presents some mathematical techniques developed to study the existence of weak solutions for a class of fluid–structure interaction problems. In particular, the interaction is between viscous, incompressible fluids and elastic, viscoelastic, or composite structures. The proofs are constructive, being based on semidiscretizing the coupled problem in time, with the Rothe method, and an operator splitting method. Then convergence of the scheme is proved by using a recent generalization of the Aubin–Lions–Simon compactness lemma, which can be useful also in other contexts.

Prof. P. Constantin in the contribution: "Regularity and Inviscid Limits in Hydrodynamic Models" discusses the problem of vanishing viscosity limit and the related low regularity bounds, uniform in viscosity. This is done for the vorticity in the Yudovich class in two-space dimensions. He also considers the problem of construction of multi-scale steady solutions of the Navier–Stokes equations with a given power-law energy spectrum (in a remarkable way this includes also the Kolmogorov K41 case). Results are not limited to the problem without boundaries, but they can be extended to any domain in three-space dimensions.

Prof. A. Kiselev in the contribution: "Small Scale Creation in Active Scalars" focuses on small-scale formation in solutions of the incompressible fluid dynamics and associated models. First, there is a review of the two-dimensional case and then the three-dimensional scenario by Hou and Luo is discussed with great detail. Examples on the surface quasi-geostrophic (SQG) equation, with emphasis on patch solutions are also studied.

It is our pleasure to thank the lecturers for their enthusiastic participation in the school, for delivering their lectures, and also for writing up the notes. We also thank all participants for their keen scientific and social contribution to the success of the school. We thank the Fondazione CIME and its scientific committee for giving us the opportunity to organize this event and especially Elvira Mascolo and Paolo Salani, Director and Scientific Secretary of the CIME Foundation, for their help in the preparation of the CIME School and of this volume. We gratefully acknowledge the Universities of Freiburg and Pisa and the CIME Foundation, for the financial support.

Pisa, Italy   Luigi C. Berselli
Freiburg, Germany   Michael Růžička
February 2020

Contents

Chapter 1
A Heuristic Approach to Convex Integration for the Euler Equations

Tristan Buckmaster and Vlad Vicol

Abstract The purpose of these lecture notes is to employ a heuristic approach in designing a convex integration scheme that produces non-conservative weak solutions to the Euler equations.

1.1 Convex Integration as a Mathematical Tool to Resolve Onsager's Conjecture

In these lecture notes, we aim to outline how a convex integration can be used to construct non-conservative weak solutions to the Euler equations:

$$\partial_t v + \operatorname{div}(v \otimes v) + \nabla p = 0,$$
$$\operatorname{div} v = 0. \tag{1.1}$$

We will restrict ourselves to considering the Euler equations on the periodic torus $\mathbb{T}^3$ for times $t \in (-1, 1)$. It is easy to check, after a simple integration by parts, that for smooth solutions to the Euler equation, the kinetic energy, defined by

$$E(t) := \frac{1}{2} \int_{\mathbb{T}^3} |v(t)|^2 \, dx,$$

is conserved. This calculation however does not hold for weak solutions. Indeed, the theory of turbulence naturally leads one to study the existence of dissipative

T. Buckmaster (✉)
Department of Mathematics, Princeton University, Princeton, NJ, USA
e-mail: buckmaster@math.princeton.edu

V. Vicol
Courant Institute for Mathematical Sciences, New York University, New York, NY, USA
e-mail: vicol@cims.nyu.edu

© The Editor(s) (if applicable) and The Author(s), under exclusive
licence to Springer Nature Switzerland AG 2020
L. C. Berselli, M. Růžička (eds.), *Progress in Mathematical Fluid Dynamics*,
Lecture Notes in Mathematics 2272, https://doi.org/10.1007/978-3-030-54899-5_1

1

weak solutions to the Euler equations. If one views the Euler equation as an inviscid limit of the Navier–Stokes equations, then formally, if one takes the inviscid limit of *turbulent* solutions, then one expects to obtain solutions to the Euler equations that dissipate kinetic energy, and are therefore necessarily weak solutions (see for example Section 2 of [5] and the references within for a more detailed discussion). The postulate of dissipation of kinetic energy at the inviscid limit is sometimes referred to in the literature as the *zeroth law of turbulence*.

In [20], Onsager famously conjectured the following dichotomy:

Conjecture 1.1 (Onsager's Conjecture)

(a) Any weak solution v belonging to the Hölder space C^θ for $\theta > \frac{1}{3}$ conserves kinetic energy.
(b) For any $\theta < \frac{1}{3}$ there exist weak solutions $v \in C^\theta$ which dissipate kinetic energy.

Part (a) was resolved by Constantin, E and Titi in [9], following a partial resolution of Eyink in [16] (see also [7, 15] for more refined results). The first result towards proving Part (b) was the construction of non-conservative L^2 weak solution to the Euler equations by Scheffer [21]. While the solutions constructed by Scheffer were non-conservative, they could not be classed a dissipative since they did not satisfy the property of non-increasing energy. The first example of dissipative weak solutions to the Euler equations was due to Shnirelman in [22] (cf. [11, 12]). Motivated in part by the convex integration scheme of Nash, employed in order to construct exotic counter-examples to the C^1 isometric embedding problem [19], De Lellis and Székelyhidi Jr. in [13, 14], made significant progress towards Part (b) by constructing dissipative Hölder $C^{\frac{1}{10}-}$ continuous weak solutions to the Euler equations. Then after a series on advancements [1–3, 10, 17], Isett resolved the conjecture in [18]. However, like the original paper of Scheffer [21], the weak solutions constructed by Isett [18] were not strictly dissipative. This technical issue was resolved in a paper by the authors in collaboration with De Lellis and Székelyhidi Jr. [4], in which the precise statement of Part (b) was proven.

Instead of considering the more difficult problem of proving Part (b), let us consider the simpler problem of constructing non-trivial, non-conservative, Hölder continuous weak solutions:

Theorem 1.1 *For some Hölder exponent β, there a non-trivial weak solution to the Euler equations $v \in C((-1, 1); C^\beta(\mathbb{T}^3))$ with compact support in time.*

The purpose of these notes is to provide an outline of how to go about constructing a convex integration scheme in order prove Theorem 1.1. The outline will track closely with the approach taken in Section 5 of the review paper [5], which itself is based on the works [2, 11, 12, 17]. In this presentation, we eschew mathematical rigor in favor rough heuristics. This will allow us to better illustrate the main ideas that go into designing a convex integration of the type pioneered by De Lellis and Székelyhidi Jr. in [13], without getting caught up in the nitty gritty technicalities that a rigorous approach entails.

1.2 The Iteration

The general strategy for proving a theorem such as Theorem 1.1 is to construct a sequence $(v_q, \mathring{R}_q)$ of solutions to the *Euler–Reynolds* system

$$\partial_t v_q + \operatorname{div}(v_q \otimes v_q) + \nabla p_q = \operatorname{div} \mathring{R}_q, \qquad \operatorname{div} v_q = 0 \qquad (1.2)$$

such that $\mathring{R}_q \to 0$ uniformly and $v_q \to v \in C^\beta$, whereby v is a non-trivial weak solution to the Euler equations (1.1) with compact support in time. The tensor $\mathring{R}_q$ is called the *Reynolds stress*, and is assumed to be symmetric and trace-free. At each inductive step, the perturbation

$$w_{q+1} = v_{q+1} - v_q$$

is designed such that v_{q+1} satisfies (1.2) with a smaller Reynolds stress $\mathring{R}_{q+1}$. It will prove helpful to split the Reynolds stress $\mathring{R}_{q+1}$ into several components.*

$$\operatorname{div} \mathring{R}_{q+1} = \underbrace{\operatorname{div}\left(w_{q+1} \otimes w_{q+1} + \mathring{R}_{q+1}\right) + \nabla(p_{q+1} - p_q)}_{\text{oscillation error}} \qquad (1.3)$$

$$+ \underbrace{\partial_t w_{q+1} + v_{q+1} \cdot \nabla w_{q+1}}_{\text{transport error}} + \underbrace{w_{q+1} \cdot \nabla v_q}_{\text{Nash error}}.$$

The Reynolds stress $\mathring{R}_{q+1}$ can be then be solved using a -1 order linear differential operator $\mathcal{R}$, defined as follows:

Definition 1.1 The operator $\mathcal{R}$ is defined on mean zero vector fields by

$$(\mathcal{R}v)^{k\ell} = (\partial_k \Delta^{-1} v^\ell + \partial_\ell \Delta^{-1} v^k - \frac{1}{2}(\delta_{k\ell} + \partial_k \partial_\ell \Delta^{-1}) \operatorname{div} \Delta^{-1} v \,.$$

The operator $\mathcal{R}$ is formally an inverse of the divergence equation, i.e. $\operatorname{div} \mathcal{R}v = v$ for any smooth, mean zero vector field v. Moreover, the matrix $\mathcal{R}v$ is symmetric and trace free.

Suppose we are given a smooth vector field $b : \mathbb{T}^3 \to \mathbb{R}^3$ and a smooth phase function $\Phi : \mathbb{T}^3 \to \mathbb{T}^3$ satisfying for all $x \in \mathbb{T}^3$ the bound

$$C^{-1} \le |\nabla\Phi(x)| \le C \,.$$

Since $\mathring{R}_{q+1}$ is a -1 order linear differential operator, then for any $\alpha \in (0, 1)$, and λ sufficiently large, we expect an estimate of the form

$$\left\| \mathcal{R}\left(a e^{i\lambda\xi \cdot \Phi(x)}\right) \right\|_{C^\alpha} \lesssim \frac{\|a\|_{C^0}}{\lambda^{1-\alpha}} + \text{error} \,,$$

where the error can be made arbitrarily small by taking λ sufficiently large. See for example [10, Lemma 2.2]) or [5, Lemma 5.6]), for a mathematically rigorous statement. The implicit constant in the above inequality depends on α. In our scheme we will take α to be sufficiently small, and thus for the matter of heuristics we will ignore the loss of λ^{α}, as well as the additional 'error', and instead pretend that we have the estimate

$$\left\| \mathcal{R} \left(a e^{i\lambda \xi \cdot \Phi(x)} \right) \right\|_{C^0} \lesssim \frac{\|a\|_{C^0}}{\lambda}. \tag{1.4}$$

Roughly, the perturbation w_{q+1} will be of the form

$$w_{q+1} = \sum_{\xi \in \Lambda} a_\xi W_{\xi, \lambda_{q+1}} \tag{1.5}$$

where Λ is a finite set of directions, the vector fields $W_{\xi, \lambda_{q+1}}$ are oscillatory *building blocks* oscillating in direction ξ, and a_ξ are coefficient functions chosen such that

$$\sum_{\xi \in \Lambda} a_\xi^2 \fint_{\mathbb{T}^3} W_{\xi, \lambda_{q+1}} \mathring{\otimes} W_{\xi, \lambda_{q+1}} = -\mathring{R}_q. \tag{1.6}$$

Here $\mathring{\otimes}$ represents the projection of the outer product onto trace free tensors. The building blocks $W_{\xi, \lambda_{q+1}}$ will oscillate at a frequency parameterized by λ_{q+1}. The cancellation (1.6) will be essential in estimated the oscillation error defined in (1.3). Let us heuristically assume that the frequencies scale geometrically

$$\lambda_q = \lambda^q \tag{1.7}$$

for some large $\lambda \in \mathbb{N}$.[1] Then for v_q to converge in $v \in C^\beta$, we roughly require

$$\|w_{q+1}\|_{C^0} \le \lambda_{q+1}^{-\beta}. \tag{1.8}$$

Recalling that $v_q = \sum_{q'=0}^q w_{q'}$, where $w_{q'}$ oscillates at frequency $\lambda_{q'} = \lambda^{q'}$; then, (1.8) roughly translates into the estimate

$$\|v_q\|_{C^1} \le \sum_{q'=0}^q \lambda_{q'}^{1-\beta} \lesssim \lambda_q^{1-\beta}, \tag{1.9}$$

---

[1] In practice is often simpler to assume that the frequencies grow super-exponentially (cf. [2, 5, 12, 17]). However, for the purpose of heuristics, geometric growth simplifies some of the calculations.

assuming that λ is chosen sufficiently large. In view of (1.8), for such a cancellation of the type (1.6) to occur, we would require the following estimate on $\mathring{R}_q$

$$\left\| \mathring{R}_q \right\|_{C^0} \leq \lambda_{q+1}^{-2\beta} . \tag{1.10}$$

Utilizing that the building blocks $W_{\xi, \lambda_{q+1}}$ oscillate at frequency λ_{q+1}, then by heuristically using an estimate of the type (1.4), it is now possible to attain a heuristic estimate on the Nash error defined in (1.3)

$$\left\| \mathcal{R}(w_{q+1} \cdot \nabla v_q) \right\|_{C^0} \lesssim \frac{\left\| w_{q+1} \right\|_{C^0} \left\| v_q \right\|_{C^1}}{\lambda_{q+1}} \lesssim \lambda_{q+1}^{-1-\beta} \lambda_q^{1-\beta} \lesssim \lambda_{q+2}^{-2\beta} \lambda^{3\beta-1} ,$$

where we used (1.8) and (1.9) in the second inequality, and (1.7) in the last inequality.

Since the Nash error forms part of the Reynolds stress error $\mathring{R}_{q+1}$, in order that (1.10) is satisfied (with q replaced by $q + 1$), we require that $\beta < \frac{1}{3}$.

1.2.1 Beltrami Flows

We are yet to define the building blocks $W_{\xi, \lambda_{q+1}}$ used in the definition of the perturbation w_{q+1}. There a number of different options depending on the goals of the convex integration schemes: *Beltrami flows*, were first utilized in the context of a convex integration scheme by De Lellis and Székelyhidi Jr. in [13]; *Mikado flows*, introduced by Daneri and Székelyhidi Jr. in [10], were essential in resolving Onsager's conjecture; *intermittent Beltrami flows*, were used in the first non-uniqueness result for weak solution to the Navier–Stokes equations [6]; and *intermittent jets*, were introduced as an improvement on intermittent Beltrami flows [8]. For the purpose of this note, we will employ Beltrami flows as our building blocks, as in [13].

A stationary divergence free vector field v is called a *Beltrami flow* if it satisfies the *Beltrami condition*:

$$\lambda(x)v(x) = \mathrm{curl}\, v(x), \qquad \lambda(x) > 0 ,$$

for all x. The function λ is called the *Beltrami coefficient*. For the purpose of these notes, we will assume that the Beltrami coefficient is a constant.

Given a Beltrami flow v, from the divergence free condition we have the following identity

$$\mathrm{div}\,(v \otimes v) = v \cdot \nabla v = \nabla \frac{|v|^2}{2} - v \times (\mathrm{curl}\, v) = \nabla \frac{|v|^2}{2} - \lambda v \times v = \nabla \frac{|v|^2}{2} .$$

In particular setting $p := \frac{|v|^2}{2}$, then (v, p) is a stationary solution to the Euler equations. Note that trivially, any linear sum of Beltrami flows with the same Beltrami coefficient λ, is itself a Beltrami flow with the Beltrami coefficient λ. This later property will be used to create a large family Beltrami flows, which will be necessary in order to achieve the cancellation (1.6).

Let us now define $W_{\xi,\lambda_{q+1}}$. We will suppose that $\xi \in \mathbb{S}^2 \cap \mathbb{Q}^3$ is such that $\lambda_{q+1}\xi \in \mathbb{Z}^3$. We define $A_\xi \in \mathbb{R}^3$ and $B_\xi \in \mathbb{C}^3$ by

$$A_\xi \cdot \xi = 0, \quad A_{-\xi} = A_\xi \quad \text{and} \quad B_\xi = \frac{1}{\sqrt{2}}\left(A_\xi + i\xi \times A_\xi\right).$$

We then observe that B_ξ satisfies the following properties

$$|B_\xi| = 1, \quad B_\xi \cdot \xi = 0, \quad i\xi \times B_\xi = B_\xi, \quad B_{-\xi} = \overline{B_\xi}.$$

Hence the vector field

$$W_{\xi,\lambda_{q+1}}(x) := B_\xi e^{i\lambda_{q+1}\xi \cdot x} \tag{1.11}$$

is $\mathbb{T}^3$ periodic (using that $\lambda_{q+1}\xi \in \mathbb{Z}^3$), divergence free, and is an eigenfunction of the curl operator with eigenvalue λ. That is, $W_{\xi,\lambda_{q+1}}$ is a complex Beltrami plane wave with Beltrami coefficient λ_{q+1}. A real valued Beltrami plane wave with Beltrami coefficient λ_{q+1} is then attained by summing $W_{\xi,\lambda_{q+1}}$ with its complex conjugate. In view of this, we define

$$W_{-\xi,\lambda_{q+1}} = \overline{W}_{\xi,\lambda_{q+1}}.$$

Then in order to ensure the right-hand-side of (1.5) is real valued, it will suffice that $\overline{a}_\xi = a_{-\xi}$; or more simply, $a_\xi = a_{-\xi}$, if we further assume the coefficients a_ξ to be real valued. Now, for the moment let us assume that the coefficients a_ξ are chosen to be real valued constants—later, we will allow dependence on x. We further suppose that Λ is a finite subset of $\mathbb{S}^2 \cap \mathbb{Q}^3$ such that $-\Lambda = \Lambda$, and moreover $\lambda_{q+1}\Lambda \subset \mathbb{Z}^3$. Then the vector field

$$W(x) = \sum_{\xi \in \Lambda} a_\xi B_\xi e^{i\lambda_{q+1}\xi \cdot x}$$

is a real-valued, divergence-free Beltrami vector field satisfying $\operatorname{curl} W = \lambda_{q+1}W$. Moreover, from the identity $B_\xi \otimes B_{-\xi} + B_{-\xi} \otimes B_\xi = 2\operatorname{Re}(B_\xi \otimes B_{-\xi}) = \operatorname{Id} - \xi \otimes \xi$, we have

$$\fint_{\mathbb{T}^3} W \otimes W \, dx = \frac{1}{2}\sum_{\xi \in \Lambda} a_\xi^2 \left(\operatorname{Id} - \xi \otimes \xi\right). \tag{1.12}$$

We refer the reader to Proposition 3.1 in [13] for additional details regarding the computations above.

1.3 Oscillation Error

In this section, we demonstrate how the Beltrami flows of the previous section can be used in order the minimize the oscillation error defined in (1.3).

In order to maintain notational consistency with previous convex integration schemes in the literature, we introduce the amplitude parameter

$$\delta_{q+1} = \lambda_{q+1}^{-2\beta}.$$

Applying a little bit of linear algebra, it is not difficult to construct a finite set $\Lambda \subset \mathbb{S}^2 \cap \mathbb{Q}^3$ and coefficient functions a_ξ for each $\xi \in \Lambda$ whose amplitude is proportional to the square root of the uniform norm of $\mathring{R}_q$, that is by (1.10) we have $a_\xi = O(\delta_{q+1}^{\frac{1}{2}})$, in such a way as to achieve the cancellation (1.6). More specifically, we define

$$a_\xi(x, t) = \delta_{q+1}^{\frac{1}{2}} \gamma_\xi \left(\mathrm{Id} - \frac{\mathring{R}_q(x, t)}{\delta_{q+1}} \right) \tag{1.13}$$

where γ_ξ are smooth functions whose domain consists of a small ball around the identity matrix within the space of symmetric matrices. We refer the reader to Lemma 3.2 in [13] (alternatively Lemma 1.3 in [2]) for the precise definition of γ_ξ. Technically, in order that the definition (1.13) makes sense, we require a slightly stronger bound than (1.10) in order to ensure that $\mathrm{Id} - \frac{\mathring{R}_q(x,t)}{\delta_{q+1}}$ lies in the image of γ_ξ. For the purpose of this note, we ignore this minor technicality.

Assuming uniform bounds on the functions γ_ξ, we obtain the following bounds on a_ξ

$$\left\| a_\xi \right\|_{C^0} \lesssim \delta_{q+1}^{\frac{1}{2}} \tag{1.14}$$

$$\left\| \nabla a_\xi \right\|_{C^0} \lesssim \delta_{q+1}^{-\frac{1}{2}} \left\| \mathring{R}_q \right\|_{C^1} \tag{1.15}$$

We now define our perturbation w_{q+1} to be

$$w_{q+1} = \sum_{\xi \in \Lambda} a_\xi W_{\xi, q+1}. \tag{1.16}$$

Then by construction, we have (1.6). Let us now compute the term $\operatorname{div}(w_{q+1} \otimes w_{q+1} + \mathring{R}_q)$ that appears in the oscillation error, defined in (1.3)

$$
\begin{aligned}
&\operatorname{div}\left(w_{q+1} \otimes w_{q+1} + \mathring{R}_q\right) \\
&\quad = \sum_{\xi \in \Lambda} \operatorname{div}\left(a_\xi^2(\operatorname{Id} - \xi \otimes \xi) + \mathring{R}_q\right) + \sum_{\xi + \xi' \neq 0,\, \xi, \xi' \in \Lambda} \operatorname{div}\left(a_\xi a_{\xi'} W_\xi \otimes W_{\xi'}\right) \\
&\quad = \nabla r_1 + \underbrace{\sum_{\xi + \xi' \neq 0,\, \xi, \xi' \in \Lambda} \left(W_\xi \otimes W_{\xi'}\right) \nabla\left(a_\xi a_{\xi'}\right)}_{:= I} \\
&\qquad + \frac{1}{2} \underbrace{\sum_{\xi + \xi' \neq 0,\, \xi, \xi' \in \Lambda} a_\xi a_{\xi'} \operatorname{div}\left(W_\xi \otimes W_{\xi'} - W_{\xi'} \otimes W_\xi\right)}_{:= II}
\end{aligned}
$$

where the pressure r_1 is implicitly defined. Applying the estimate (1.4) with $\Phi(x) = x$ and ξ replaced by $\xi + \xi'$, together with (1.14) and (1.15), we obtain

$$
\begin{aligned}
\|\mathcal{R}(I)\|_{C^0} &\lesssim \frac{1}{\lambda_{q+1}} \sum_{\xi + \xi' \neq 0,\, \xi, \xi' \in \Lambda} \left\|\nabla(a_\xi a_{\xi'})\right\|_{C^0} \\
&\lesssim \frac{1}{\lambda_{q+1}} \sum_{\xi + \xi' \neq 0,\, \xi, \xi' \in \Lambda} \left\|a_\xi\right\|_{C^0} \left\|\nabla a_{\xi'}\right\|_{C^0} \\
&\lesssim \frac{1}{\lambda_{q+1}} \left\|\mathring{R}_q\right\|_{C^1}.
\end{aligned}
$$

Thus, this error can be made small by assuming λ_{q+1} to be sufficiently large. Now consider II. We write

$$
II = \nabla r_2 - \frac{1}{2} \underbrace{\sum_{\xi + \xi' \neq 0,\, \xi, \xi' \in \Lambda} \nabla(a_\xi a_{\xi'})\left(W_\xi \cdot W_{\xi'}\right)}_{:= III}
$$

with the pressure r_2 again being implicitly defined. Then III can be estimated in the same manner as I. Hence, setting $p_{q+1} = p_q - r_1 - r_2$, we are able to attain suitable bounds on the contribution of the oscillation error to $\mathring{R}_{q+1}$.

An issue with the definition (1.16), is the vector field w_{q+1} is not necessarily divergence free. To fix this, we relabel the right-hand-side of (1.16) to be the *principal perturbation* $w_{q+1}^{(p)}$, i.e.

$$
w_{q+1}^{(p)} := \sum_{\xi \in \Lambda} a_\xi W_{\xi, q+1}. \tag{1.17}
$$

We then define a *corrector* $w_{q+1}^{(c)}$ by

$$w_{q+1}^{(p)} := \frac{1}{\lambda_{q+1}} \sum_{\xi \in \Lambda} \nabla a_\xi \times W_{\xi,q+1} . \tag{1.18}$$

Finally, defining

$$w_{q+1} = w_{q+1}^{(p)} + w_{q+1}^{(c)} , \tag{1.19}$$

then a simple calculation yields

$$w_{q+1} = \frac{1}{\lambda_{q+1}} \sum_{\xi \in \Lambda} \mathrm{curl} \left(a_\xi W_{\xi,q+1} \right) ,$$

from which it follows that w_{q+1} is divergence free. Moreover, assuming λ_{q+1} is sufficiently large, the corrector $w_{q+1}^{(c)}$ is small, and hence can be made to have a suitably small contribution to the Reynolds stress $\mathring{R}_{q+1}$ resulting from the perturbation defined in (1.19).

1.4 Transport Error

We now consider the transport error

$$\underbrace{(\partial_t + v_q \cdot \nabla)}_{D_t} w_{q+1} ,$$

defined in (1.3), where here D_t represents the *material derivative* associated with v_q. Ignoring the contribution of the corrector to the transport error, by definition (1.19), applying (1.4), we heuristically attain

$$\left\| \mathcal{R}(D_t w_{q+1}^{(p)}) \right\|_{C^0} \lesssim \frac{1}{\lambda_{q+1}} \sum_{\xi \in \Lambda} \left\| D_t a_\xi \right\|_{C^0} \left\| W_{\xi,q+1} \right\|_{C^0}$$

$$+ \frac{1}{\lambda_{q+1}} \sum_{\xi \in \Lambda} \left\| a_\xi \right\|_{C^0} \left\| v_q \cdot \nabla W_{\xi,q+1} \right\|_{C^0} .$$

The second term is unfortunately not small, since $\nabla W_{\xi,q+1} = O(\lambda_{q+1})$. To rectify this issue we will replace $W_{\xi,q+1}$ in the ansatz (1.17) with

$$W_{(\xi)} = W_{\xi,j,q+1} = W_{\xi,q+1} \circ \Phi_j$$

where Φ_j are phase functions solving the transport equation

$$D_t \Phi_j \equiv 0, \qquad \Phi(x, j\ell) = x$$

for some small parameter $\ell > 0$ to be chosen later and $j \in \mathbb{Z}$. With this definition, we have

$$D_t W_{(\xi)} \equiv 0 .$$

The trade-off with using $W_{(\xi)}$ in place of $W_{\xi,q+1}$, is that $W_{(\xi)}$ is no longer an exact eigenfunction of curl. Let us write

$$W_{(\xi)} = B_\xi e^{i\lambda_{q+1}\xi \cdot \Phi_j} = \underbrace{e^{i\lambda_{q+1}\xi \cdot (\Phi_j - x)}}_{:= \Phi_{(\xi)}} W_\xi$$

then

$$\operatorname{curl} W_{(\xi)} = \lambda_{q+1} W_\xi + \nabla \phi_{(\xi)} \times W_{\xi,q+1} . \tag{1.20}$$

Thus in order to quantity how well $W_{(\xi)}$ approximates an eigenfunction of curl, we need to estimate $\nabla \phi_{(\xi)}$. By standard transport estimates we obtain

$$\left\| \nabla \Phi_j - \operatorname{Id} \right\|_{C^0} \leq \exp \left(\int_{j\ell}^t \left\| v_q(s) \right\|_{C^1} ds \right) .$$

In particular, if $|t - j\ell| \leq \left\| v_q \right\|_{C^1}^{-1}$, we have

$$\left\| \nabla \Phi_j - \operatorname{Id} \right\|_{C^0} \lesssim |t - j\ell| \left\| v_q \right\|_{C^1} .$$

From which we deduce

$$\left\| \nabla \phi_{(\xi)} \right\|_{C^0} \lesssim \lambda_{q+1} |t - j\ell| \left\| v_q \right\|_{C^1} . \tag{1.21}$$

Thus $W_{(\xi)}$ is Beltrami like so long that $|t - j\ell|$ is suitably small. To achieve this, we partition time, and replace a_ξ with new coefficient functions $a_{(\xi)}$ with small temporal support. We introduce cut-off functions $\chi_j : (-1, 1) \to \mathbb{R}$ with support contained in the interval $(\ell(j-2), \ell(j+2))$, such that the squares χ_j^2 form a partition of unity, i.e.

$$\sum_j \chi_j^2 \equiv 1 .$$

In place of Λ, we will require two disjoint finite subsets $\Lambda^{(0)}, \Lambda^{(1)} \subset \mathbb{S}^2 \cap \mathbb{Q}^3$. The set $\Lambda^{(0)}$ can be taken to be Λ, and $\Lambda^{(1)}$ can be defined in terms of a rational rotation of Λ. Similarly in place of the family of smooth functions γ_ξ, we will require two families of smooth functions $\{\gamma_\xi^{(0)} \mid \xi \in \lambda^{(0)}\}$ and $\{\gamma_\xi^{(1)} \mid \xi \in \lambda^{(1)}\}$. Again, we refer to Lemma 3.2 in [13] (alternatively Lemma 1.3 in [2]) for the precise definitions of the sets $\Lambda^{(j)}$ and functions $\gamma_\xi^{(j)}$. We then define

$$a_{(\xi)} = \delta_{q+1}^{\frac{1}{2}} \chi_j \gamma_\xi^{(j)} \left(\mathrm{Id} - \frac{\mathring{R}_q(x,t)}{\delta_{q+1}} \right),$$

where by an abuse of notation we write $\gamma_\xi^{(j)} = \gamma_\xi^{(j \bmod 2)}$. With these definitions, we replace the definitions $w_{q+1}^{(p)}$ and $w_{q+1}^{(c)}$ given in (1.17) and (1.18) respectively with the new definitions

$$
\begin{aligned}
w_{q+1}^{(p)} &:= \sum_j \sum_{\xi \in \Lambda_j} a_{(\xi)} W_{(\xi)} \\
w_{q+1}^{(c)} &:= \frac{1}{\lambda_{q+1}} \sum_j \sum_{\xi \in \Lambda_j} \nabla(a_{(\xi)} \phi_{(\xi)}) W_{(\xi)}
\end{aligned}
\tag{1.22}
$$

where we use the notation $\Lambda_j = \Lambda_{j \bmod 2}$. The functions $\gamma_\xi^{(j)}$ are again defined in such a way that we achieve a cancellation analogous to (1.6). More precisely, in place of (1.6), we have

$$\sum_j \sum_{\xi \in \Lambda_j} a_\xi^2 \fint_{\mathbb{T}^3} W_{\xi,\lambda_{q+1}} \mathring{\otimes} W_{\xi,\lambda_{q+1}} = -\mathring{R}_q.$$

The principal reason for introducing the two families $\Lambda^{(0)}$, $\Lambda^{(1)}$ was to reduce the interactions between the oscillatory Beltrami waves across the neighboring temporal regions where the cut-off functions χ_j overlap.

Due to the small prefactor in the definition of $w_{q+1}^{(c)}$, the term $D_t w_{q+1}^{(p)}$ will be the the main contribution of $D_t w_{q+1}$ to the transport error. Hence, in order to estimate the transport error, we will need bounds on

$$D_t a_{(\xi)} W_{(\xi)} = (D_t a_{(\xi)}) W_{(\xi)}.$$

By definition

$$\|D_t \chi_j\|_{C^0} \lesssim \ell^{-1}.$$

The material derivative falling on the cut-off is expected to produce the main contribution to the transport error. Conversely, owing to the calculation (1.20), the main contribution to the new oscillation error associated with the new perturbation definition (1.22) occurs when derivatives fall on $\phi_{(\xi)}$. Recalling (1.21), we have

$$\|\nabla \phi_{(\xi)}\|_{C^0} \lesssim \lambda_{q+1} |t - j\ell| \|v_q\|_{C^1}$$

Thus in order to balance the transport and oscillation error, it is necessary to optimize our choice of ℓ. Making the appropriate choice, we can simultaneously obtain effective bounds on the oscillation, transport and Nash errors in (1.3) in order to ensure that (1.10) holds with $q + 1$ replacing q.[2]

1.5   Mollification and Loss of Derivative Problem

Recall that $(v_q, \mathring{R}_q)$ are defined inductively. In order to ensure convergence to a solution, one needs to inductively keep track of estimates on $(v_q, \mathring{R}_q)$. As the current scheme is currently defined above, the definition of $\mathring{R}_q$ involves derivatives of $\mathring{R}_{q-1}$ (for example when a derivative falls on $a_{(\xi)}$ in the oscillation error), which in turn involves higher order derivatives on $\mathring{R}_{q-2}$, and so forth. Thus in order for the scheme to close, one would have to keep track of estimates on infinitely many derivatives of $(v_q, \mathring{R}_q)$. To avoid this *loss of derivative problem*, we introduce an addition step where we replace $(v_q, \mathring{R}_q)$ with the mollified $(v_\ell, \mathring{R}_\ell)$ defined by

$$v_\ell = (v_q *_x \psi_\ell)) *_t \varphi_\ell, \qquad \text{and} \qquad \mathring{R}_\ell = (\mathring{R}_q *_x \psi_\ell)) *_t \varphi_\ell$$

where ψ_ℓ and φ_ℓ are standard space and time mollifiers respectively. Then we have

$$\partial v_\ell + \text{div}\,(v_\ell \otimes v_\ell) + \nabla p_\ell = \text{div}\left(\mathring{R}_\ell + \underbrace{v_\ell \mathring{\otimes} v_\ell - ((v_q \mathring{\otimes} v_q) *_x \psi_\ell)) *_t}_{R_{\text{commutator}}}\right).$$

The new error $R_{\text{commutator}}$ can be made small by assuming ℓ to be sufficiently small. With $(v_\ell, \mathring{R}_\ell)$ defined above, in the definition of w_{q+1} described above, we replace all references of v_q and $\mathring{R}_q$ with v_ℓ and $\mathring{R}_\ell$. Then the new velocity field v_{q+1} is defined by

$$v_{q+1} := v_\ell + w_{q+1}.$$

With this additional mollification step, we no longer need to keep track of infinitely many derivatives of $(v_q, \mathring{R}_q)$, indeed it will suffice to keep track of C^0 and C^1 estimates on $(v_q, \mathring{R}_q)$.

---

[2]It should be noted however that in order for the scheme described here to close, one should replace the geometric growth of frequencies λ_q described in (1.7) with superexponential growth. A scheme involving geometric growth of frequencies is slightly more delicate to describe.

1.6 Compact Support in Time

We close out these notes by outlining an argument in order to achieve non-trivial weak solutions with compact support in time.

In order to ensure $v = \lim_q v_q$ has compact support in time, we inductively assume that

$$\operatorname{supp}_t v_q \cup \operatorname{supp}_t \mathring{R}_q \subset \left[-\frac{1}{2} + 2^{-q-2}, \frac{1}{2} - 2^{-q-2}\right]. \qquad (1.23)$$

The mollification step, will increase the temporal support, assuming the temporal mollifier φ_ℓ is suitably defined, we have

$$\operatorname{supp}_t v_\ell \cup \operatorname{supp}_t \mathring{R}_\ell \subset \left[-\frac{1}{2} + 2^{-q-2} - \ell, \frac{1}{2} - 2^{-q-2} + \ell\right].$$

Then in order to correct the Reynolds stress R_ℓ, we need only sum j in the definition (1.22), for j satisfying

$$\operatorname{supp}_t \chi_j \subset \left[-\frac{1}{2} + 2^{-q-2} - 4\ell, \frac{1}{2} - 2^{-q-2} + 4\ell\right].$$

Hence choosing ℓ sufficiently small we have

$$\operatorname{supp}_t v_{q+1} \cup \operatorname{supp}_t \mathring{R}_{q+1} \subset \left[-\frac{1}{2} + 2^{-q-2} - 4\ell, \frac{1}{2} - 2^{-q-2} + 4\ell\right]$$

$$\subset \left[-\frac{1}{2} + 2^{-q-3}, \frac{1}{2} - 2^{-q-3}\right],$$

and thus we attain (1.23) with $q + 1$ replacing q. Hence for $v = \lim_q v_q$ we have

$$\operatorname{supp}_t v \subset \left[-\frac{1}{2}, \frac{1}{2}\right].$$

Acknowledgments T.B. was supported by the NSF grant DMS-1900149 and a Simons Foundation Mathematical and Physical Sciences Collaborative Grant. V.V. was supported by the NSF grant DMS-1911413.

References

1. T. Buckmaster, Onsager's conjecture almost everywhere in time. Commun. Math. Phys. **333**(3), 1175–1198 (2015)
2. T. Buckmaster, C. De Lellis, P. Isett, L. Székelyhidi, Jr., Anomalous dissipation for 1/5-Hölder Euler flows. Ann. Math. **182**(1), 127–172 (2015)

3. T. Buckmaster, C. De Lellis, L. Székelyhidi, Jr., Dissipative Euler flows with Onsager-critical spatial regularity. Commun. Pure Appl. Math. **69**(9), 1613–1670 (2016)
4. T. Buckmaster, C. De Lellis, L. Székelyhidi Jr., V. Vicol, Onsager's conjecture for admissible weak solutions. Commun. Pure Appl. Math. **72**(2), 227–448 (2019)
5. T. Buckmaster, V. Vicol, *Convex Integration and Phenomenologies in Turbulence* (2019)
6. T. Buckmaster, V. Vicol, Nonuniqueness of weak solutions to the Navier–Stokes equation. Ann. Math. **189**(1), 101–144 (2019)
7. A. Cheskidov, P. Constantin, S. Friedlander, R. Shvydkoy, Energy conservation and Onsager's conjecture for the Euler equations. Nonlinearity **21**(6), 1233–1252 (2008)
8. M. Colombo, C. De Lellis, L. De Rosa, Ill-posedness of Leray solutions for the ipodissipative Navier–Stokes equations (2017). arXiv:1708.05666
9. P. Constantin, E. Weinan, E. Titi, Onsager's conjecture on the energy conservation for solutions of Euler's equation. Commun. Math. Phys. **165**(1), 207–209 (1994)
10. S. Daneri, L. Székelyhidi Jr., Non-uniqueness and h-principle for Hölder-continuous weak solutions of the Euler equations. Arch. Ration. Mech. Anal. **224**(2), 471–514 (2017)
11. C. De Lellis, L. Székelyhidi, Jr., The Euler equations as a differential inclusion. Ann. Math. (2) **170**(3), 1417–1436 (2009)
12. C. De Lellis, L. Székelyhidi, Jr., On admissibility criteria for weak solutions of the Euler equations. Arch. Ration. Mech. Anal. **195**(1), 225–260 (2010)
13. C. De Lellis, L. Székelyhidi, Jr., Dissipative continuous Euler flows. Invent. Math. **193**(2), 377–407 (2013)
14. C. De Lellis, L. Székelyhidi, Jr., On h-principle and Onsager's conjecture. Eur. Math. Soc. Newsl. **95**, 19–24 (2015)
15. J. Duchon, R. Robert, Inertial energy dissipation for weak solutions of incompressible Euler and Navier–Stokes equations. Nonlinearity **13**(1), 249 (2000)
16. G. Eyink, Energy dissipation without viscosity in ideal hydrodynamics I. Fourier analysis and local energy transfer. Phys. D: Nonlinear Phenom. **78**(3–4), 222–240 (1994)
17. P. Isett, Hölder continuous Euler flows in three dimensions with compact support in time (2012). arXiv:1211.4065
18. P. Isett, A proof of Onsager's conjecture. Ann. Math. **188**(3), 871–963 (2018)
19. J. Nash, C^1 isometric imbeddings. Ann. Math. 383–396 (1954)
20. L. Onsager, Statistical hydrodynamics. Nuovo Cimento (9), 6 (Supplemento, 2 (Convegno Internazionale di Meccanica Statistica)), 279–287 (1949)
21. V. Scheffer, An inviscid flow with compact support in space-time. J. Geom. Anal. **3**(4), 343–401 (1993)
22. A. Shnirelman, Weak solutions with decreasing energy of incompressible Euler equations. Commun. Math. Phys. **210**(3), 541–603 (2000)

Chapter 2
Fluid-Structure Interaction with Incompressible Fluids

Sunčica Čanić

Abstract These lecture notes cover a series of three two-hour lectures on fluid-structure interaction involving incompressible, viscous fluids, presented as the CIME Summer Workshop entitled "Progress in Mathematical Fluid Dynamics", held in Cetraro, Italy, in June 2019. They are intended for graduate students and postdocs with interest in mathematical fluid dynamics. The goal was to present certain mathematical techniques, developed within the past 6 years, to study existence of weak solutions for a class of fluid-structure interaction problems between viscous, incompressible fluids and elastic, viscoelastic, or composite structures. The resulting problem is a nonlinear moving-boundary problem with a strong geometric nonlinearity associated with the fluid domain motion. The existence proof is constructive. It is based on semidiscretizing the coupled problem in time, also known as Rothe's method, and then using a Lie operator splitting strategy to define fluid and structure subproblems, which communicate via the initial data. The Lie operator splitting is designed in just the right way so that the energy balance at the discrete level approximates well that of the coupled, continuous problem. To prove convergence of approximating sequences to a weak solution of the coupled problem, a recent generalization of the Aubin–Lions–Simon compactness lemma to problems on moving domains is used. The methodology presented here served as a basis for the construction of several loosely coupled partitioned schemes for solving fluid-structure interaction problems.

2.1 Introduction and Literature Review

The goal of these lecture notes is to summarize the main ideas, recently developed by B. Muha and S. Čanić in a series of papers [19, 62–67], to study existence of weak solutions to a class of fluid-structure interaction (FSI) problems involving viscous,

S. Čanić (✉)
Department of Mathematics, University of California, Berkeley, CA, USA
e-mail: canics@berkeley.edu

© The Editor(s) (if applicable) and The Author(s), under exclusive licence to Springer Nature Switzerland AG 2020
L. C. Berselli, M. Růžička (eds.), *Progress in Mathematical Fluid Dynamics*, Lecture Notes in Mathematics 2272, https://doi.org/10.1007/978-3-030-54899-5_2

incompressible fluids and elastic structures. The main novelty of this body of work is the development of mathematical techniques of *constructive existence proofs* based on Rothe's method, in the scenarios when the structure location is not known a priori but is one of the unknowns in the problem. The methodology presented here is rather robust; the construction of approximate solutions developed here has since been used in a number of different existence proofs, and also as a basis for the development of new partitioned, loosely coupled schemes for FSI problems [8–14, 16–18, 59].

Although the development of numerical methods for fluid-structure interaction problems started around 40 years ago (see e.g., [2–4, 9, 10, 17, 25, 32, 33, 37, 39, 42, 47, 48, 51, 56, 69–71, 73, 77] and the references therein), the development of existence theory started almost 20 years later. The main difficulties in the analysis of this class of problems are their multi-physics nature, resulting in mathematical problems of mixed (hyperbolic-parabolic) type, and the strong nonlinearities in the problem, which include the geometric nonlinearity when the fluid and structure are coupled across the "current" interface, not known a priori.

The first existence results addressed problems in which the structures were completely immersed in the fluid, and the structure was considered to be either a rigid body, or described by a finite number of modal functions. See e.g., [7, 24, 28, 30, 31, 38, 41, 74], and the references therein. Analysis of existence of FSI solutions involving *elastic structures* interacting with the flow of a $2D$ or $3D$ incompressible, viscous fluid started in the early 2000s. First, the coupling between the fluid and structure was assumed across a fixed fluid-structure interface, called *linear coupling*, as in [5, 6, 19, 35, 53], and then extended to problems with *nonlinear coupling* in [18, 21–23, 26, 27, 29, 45, 46, 54, 57, 58, 62–66]. More precisely, concerning nonlinear FSI models, the first FSI existence result, locally in time, was obtained by DaVega [29], where a **strong** solution for an interaction between an incompressible, viscous fluid in $2D$ and a $1D$ viscoelastic string was studied, assuming periodic boundary conditions. This result was extended by Lequeurre in [58], where the existence of a unique, local in time, strong solution for arbitrarily large data, and the existence of a global strong solution for small data, was proved in the case when the structure is modeled as a clamped viscoelastic beam. D. Coutand and S. Shkoller proved the existence, locally in time, of a unique, regular solution for an interaction between a viscous, incompressible fluid in $3D$ and a $3D$ structure, immersed in the fluid, where the structure was modeled by the equations of linear [26], or quasi-linear [27] elasticity. In the case when the structure (solid) is modeled by a linear wave equation, I. Kukavica et al. proved the existence, locally in time, of a strong solution, assuming lower regularity for the initial data [50, 54]. A similar result for compressible flows can be found in [52]. In [72] Raymod et al. considered a FSI problem between a linear elastic solid immersed in an incompressible viscous fluid, and proved the existence and uniqueness of a strong solution. Most of the above mentioned existence results for strong solutions are local in time. In [49] a global existence result for small data was obtained by Ignatova et al. for a moving boundary FSI problem involving a damped linear wave equation with some additional damping terms in the coupling conditions, showing exponential decay in time of the solution. In the case when the structure is modeled as a $2D$ elastic shell interacting with a viscous, incompressible fluid in $3D$, the existence, locally in time, of a unique regular solution was proved by Shkoller et al. in [22, 23]. We mention that the works of Shkoller et al., and Kukavica at al. were obtained in the context of Lagrangian coordinates, which were used for both the structure and fluid subproblems.

In the context of **weak** solutions, the first existence results came out in 2005 when Chambolle et al. showed the existence of a weak solution for a FSI problem between a $3D$ incompressible, viscous fluid and a $2D$ viscoelastic plate in [21]. Grandmont improved this result in [46] to hold for a $2D$ elastic plate. A constructive existence proof for the interaction between an incompressible, viscous fluid and a linearly elastic Koiter shell with transverse displacement was designed in [62].

The first constructive existence proof for a moving boundary problem was presented by Ladyzhenskaya in 1970, see [55], where the interaction between an incompressible, viscous fluid and a *given moving boundary* was constructed using a time-discretization approach, known as Rothe's method, assuming high regularity of the given interface. In 2013, Muha and Čanić designed a Rothe's-type method in the context of moving boundaries whose location is *not known a priori* [62]. Their method was then extended to a larger class of problems [18, 19, 64–66], as we describe below.

The focus in these lecture notes is on the constructive methodology developed by Muha and Čanić in [62], combined with a generalization of the Aubin–Lions– Simon compactness result, published in 2019 in [67], to obtain existence of weak solutions to nonlinear FSI problems. The techniques are presented in the context of the simplest, benchmark FSI problem, which embodies the main difficulties associated with studying this class of problems. The benchmark problem describes the interaction between the time-dependent flow of a viscous, incompressible fluid, modeled by the 2D Navier–Stokes equations in a 2D "cylinder" (rectangle) interacting with (thin) elastic lateral walls modeled by the linearly elastic Koiter shell equations. In this problem, the thin elastic structure coincides with the fluid-structure interface, and the flow is driven by the time-dependent inlet and outlet dynamic pressure data. The fluid and structure are coupled at the current location of the (moving) fluid-structure interface via two sets of coupling conditions describing continuity of fluid and structure velocities (no-slip) and balance of contact forces. This is known as a *two-way coupling* since both the fluid and structure "feel" each other through the exchange of information (energy) between the two. The fluid flow "feels" the structure through the contact forces exerted by the structure onto the fluid during the elastic structure motion and the no-slip condition, while the structure elastodynamics is driven by the jump in the contact forces (traction) across the interface. Assuming that external forcing onto the structure is zero, the structure elastodynamics is entirely driven by the normal stress (traction) exerted by the fluid onto the structure. The fact that the location of the fluid domain boundary is not known a priori but it depends on one of the unknowns of the problem, i.e., on the fluid-structure interface location (which in turn depends on the fluid forcing), gives rise to a strong geometric nonlinearity, which is one of the main difficulties in the analysis of this class of problems. The benchmark problem discussed in these notes exemplifies these difficulties, and the constructive existence proof presented here contains all the ingredients necessary for the development of a robust methodology, applicable to other scenarios. These include extensions to FSI involving 3D incompressible, viscous fluid flows [63], FSI with a nonlinearly elastic Koiter shell [65], FSI with a multi-layered thin-thick linearly elastic structure [64], FSI with the Navier-slip condition [66], and FSI with mesh-supported shells [19].

The proof of the existence of a weak solution of Leray type, presented in these notes, is based on semi-discretizing the continuous, coupled problem in time as in Rothe's method, and splitting the semi-discretized problem using the Lie operator splitting strategy [43]. The problem is split into two sub-problems, one defined by the structure equation, and one by the fluid equations. The key idea is to split the coupling conditions in just the right way so that the resulting, semi-discretized, split problem approximates well the energy exchange between the fluid and structure at the continuous level. With such a splitting, uniform energy estimates in time can be obtained to show that the approximate solutions, obtained for each fixed time step Δt, are uniformly bounded in Δt. This gives rise to weakly- and weakly*-convergent subsequences (in the corresponding topologies), as $\Delta t \rightarrow 0$. The goal is to show that the limits of these subsequences satisfy the weak formulation of the coupled, continuous problem. Since the coupled problem is nonlinear, a compactness argument needs to be used in order to show strong convergence of the subsequences, and pass to the limit in the weak formulations of approximate problems. Passing to the limit in the weak formulations is highly non-trivial, since each semi-discretized problem is defined on a different fluid domain. A generalization of the classical Aubin–Lions–Simon compactness result to problems on moving domains, obtained in [67], is used in these notes to obtain strongly convergent sub-sequences of approximate functions. This is, however, still not sufficient to pass to the limit in the weak formulations, since the fluid velocity test function depend on the fluid domain as well. A "trick" similar to that introduced by Chambole et al. in [21] is used in these notes to construct "appropriate" test functions for approximate problems, which converge uniformly in a strong topology to the test functions of the continuous problem, thereby allowing passage to the limit in the weak formulations, and proving existence of a weak solution.

The existence result is global in time in the sense that it holds until a possible contact between the elastic structures. Recently, Grandmont and Hillairet showed that contact between a rigid bottom of a fluid container, and a *viscoelastic beam*, is not possible in finite time [44]. The finite-time contact involving elastic structures interacting with an incompressible, viscous fluid, as studied in these notes, is still open. For more details about the state-of-the-art in the literature related to finite time contact in FSI see [15].

The existence result presented in these lectures was obtained in collaboration with Prof. Boris Muha of the University of Zagreb in Croatia. The author would also like to acknowledge PhD student Marija Galić of the University of Zagreb for her thoughtful contributions, and the PhD students at UC Berkeley, Jeffrey Kuan and Mitchell Taylor, for their careful reading of the manuscript and for their useful comments and suggestions.

2.2 Model Description

To fix ideas, we consider a "benchmark" FSI problem in 2D, which exemplifies the main difficulties associated with studying this class of FSI problem. The FSI problem consists of a Koiter shell interacting with the time-dependent 2D flow of a viscous, incompressible, Newtonian fluid. The fluid domain is a 2D rectangle with the top and bottom boundary elastic, corresponding to the lateral wall of the 2D cylinder. The flow is driven by the time-dependent inlet and outlet dynamic pressure data prescribed at the left and right boundary of the 2D cylinder. The elastodynamics of the lateral wall is modeled by the cylindrical Koiter shell equations. To simplify matters, we will be assuming that only the transverse, i.e., radial displacement of the Koiter shell is different from zero. While this is an assumption that has been almost exclusively used in the FSI existence results involving thin structures, two recent manuscripts have allowed all three components of the structure displacement to be different from zero [19, 66]. This gives rise to significant technical difficulties related to possible geometric degeneracies of the fluid domain due to the non-zero tangential displacement, and is the main reason why the existence results in [19, 66] are not global, namely, they hold until the time when the fluid domain degeneracy occurs. Resolving these issues is still an open problem.

Since we will be imposing the inlet and outlet data a to be symmetric with respect to the axis of symmetry of the 2D cylinder, we will be working on the upper half of the fluid domain and impose the symmetry boundary condition along the axis of symmetry, which is now serving as the bottom boundary of the fluid domain. Thus, the reference domain is $(0, L) \times (0, R)$ with the lateral (top) boundary given by $(0, L) \times \{R\}$. The spatial coordinates will be denoted by:

$$x = (z, r).$$

2.2.1 The Structure and Fluid Equations

The Structure Model The lateral wall of the 2D cylinder is assumed elastic and with small thickness $h \ll 1$. The reference configuration of the elastic lateral wall will be denoted by Γ (see Fig. 2.1):

$$\Gamma = (0, L) \times \{R\}.$$

Γ corresponds to the middle surface of an elastic shell of thickness h.

Fig. 2.1 A sketch of the fluid
domain and boundary

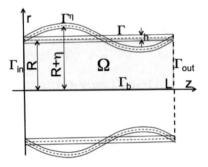

Under the fluid loading, and possibly some external loading (assumed to be zero in these notes), the cylinder will deform. We denote by $\eta = \eta(t, z)$ the displacement of the cylinder from the reference configuration Γ, and use $\Gamma^\eta(t)$ to denote the location of the cylinder lateral boundary at time t.

The elastic properties of the cylinder's lateral wall will be described by an operator $\mathcal{L}_e$, so that the elastodynamics problem, in Lagrangian formulation, can be written as:

$$\rho_s h \partial_{tt} \eta + \mathcal{L}_e \eta = \mathbf{f}, \quad \text{on } \Gamma, \ t \in (0, T), \tag{2.2.1}$$

where ρ_s is the structure density, h is the thin structure thickness, and $\mathbf{f}$ is the loading (force density) driving the motion of the elastic structure. Operator $\mathcal{L}_e$ is a *linear, continuous, positive-definite, coercive operator* on a Hilbert space χ.

Example (Cylindrical Koiter Shell) A cylindrical Koiter shell allowing only radial component of displacement to be different from zero:

$$\eta = \eta \mathbf{e}_r,$$

where $\mathbf{e}_r$ is the unit vector pointing in the vertical/radial direction of the cylinder Γ, can be written in the form (2.2.1). In that case

$$\Gamma^\eta(t) = \{(z, r) \in \mathbb{R}^2 : r = R + \eta(t, z), \ z \in (0, L)\},$$

and the cylindrical Koiter shell equations in differential form can be written as [62]:

$$\rho_s h \partial_t^2 \eta + C_0 \eta - C_1 \partial_z^2 \eta + C_2 \partial_z^4 \eta = f, \tag{2.2.2}$$

where

$$C_0 = \frac{hE}{R^2(1 - \sigma^2)} \left(1 + \frac{h^2}{12R^2}\right), \ C_1 = \frac{h^3}{6} \frac{E\sigma}{R^2(1 - \sigma^2)}, \ C_2 = \frac{h^3}{12} \frac{E}{1 - \sigma^2}.$$

Here E and σ are the Young's modulus of elasticity and Poisson ratio, respectively, describing the elastic properties of the shell, and f is the radial component of the force density $f = \mathbf{f} \cdot \mathbf{e}_r$. All the constants are assumed to be strictly greater than zero. We will be considering a clamped Koiter shell:

$$\eta(0) = \partial_z \eta(0) = \eta(L) = \partial_z \eta(L) = 0. \tag{2.2.3}$$

In this case the space $\chi = H_0^2(\Gamma)$, and the operator $\mathcal{L}_e$ is defined by:

$$\mathcal{L}_e \eta := C_0 \eta - C_1 \partial_z^2 \eta + C_2 \partial_z^4 \eta. \tag{2.2.4}$$

Throughout the rest of the manuscript we will be working with the cylindrical Koiter shell assuming only radial displacement to be different from zero.

The Fluid Model The fluid flow is modeled by the Navier–Stokes equations for an incompressible, viscous fluid:

$$\left. \begin{array}{r} \rho_f \big(\partial_t \mathbf{u} + (\mathbf{u} \cdot \nabla) \mathbf{u} \big) = \nabla \cdot \sigma \\ \nabla \cdot \mathbf{u} = 0 \end{array} \right\} \text{ in } \Omega^\eta(t), \ t \in (0, T), \tag{2.2.5}$$

where σ is the Cauchy stress tensor, ρ_f is the fluid density, and $\mathbf{u} = \mathbf{u}(x, t) = (u_z, u_r)$ is the fluid velocity. We will be working with Newtonian fluids, in which case

$$\sigma = -p\mathbf{I} + 2\mu \mathbf{D}(\mathbf{u}),$$

where μ is the dynamic viscosity coefficient, and $\mathbf{D}(\mathbf{u}) = \frac{1}{2}(\nabla \mathbf{u} + \nabla^\tau \mathbf{u})$ is the symmetrized gradient of $\mathbf{u}$.

The equations are defined on a moving domain $\Omega^\eta(t)$, which is defined by

$$\Omega^\eta(t) = \{(z, r) \in \mathbb{R}^2 : z \in (\Gamma), \ r \in (0, R + \eta(t, z)\}.$$

The flow is driven by the inlet and outlet dynamic pressure data, and the inlet and outlet flow is normal to the inlet and outlet boundary $\Gamma_{in} = \{0\} \times (0, R)$ and $\Gamma_{out} = \{L\} \times (0, R)$ (it is symmetric with respect to the axis of symmetry of the cylinder):

$$\left. \begin{array}{r} p + \dfrac{\rho_f}{2} |u|^2 = P_{in/out}(t), \\ u_r = 0, \end{array} \right\} \text{ on } \Gamma_{in/out}, \tag{2.2.6}$$

where $P_{in/out} \in L_{loc}^2(0, \infty)$ are given. At the bottom boundary $\Gamma_b = \Gamma \times \{0\}$ the symmetry boundary conditions are prescribed:

$$u_r = \partial_r u_z = 0, \quad \text{on } \Gamma_b. \tag{2.2.7}$$

Notice again that the fluid domain $\Omega^\eta(t)$ is not known a priori, since its location depends on one of the unknowns in the problem, i.e., the location of the later boundary η. In turn, the location of the lateral boundary depends on the fluid loading, giving rise to a geometric nonlinearity, which presents one of the main difficulties in studying this coupled fluid-structure interaction problem.

2.2.2   The Coupling

The fluid flow influences the motion of the structure through traction forces, i.e., by the normal stress exerted onto the structure at $\Gamma^\eta(t)$, while the structure influences the fluid through its inertial and elastic forces associated with the kinetic and potential energy of the shell. The shell responds to the fluid loading by stretching/recoil (associated with the change in the metric tensor of Γ) and by bending (associated with the change in the curvature tensor of Γ). Additionally, the fluid and structure "feel" each other through the coupling of kinematic quantities, namely, the fluid and structure velocities. We will be assuming that at the fluid-structure interface $\Gamma^\eta(t)$ the fluid and structure velocities satisfy the no-slip condition in the sense that the trace of the fluid velocity on $\Gamma^\eta(t)$ is the same as the velocity of $\Gamma^\eta(t)$ itself.

Thus, two coupling conditions, i.e., boundary conditions on $\Gamma^\eta(t)$, need to be prescribed to give rise to a well-defined problem: the kinematic coupling condition describing the coupling of kinematic quantities, and the dynamic coupling condition describing the dynamic balance of forces:

$$\partial_t \boldsymbol{\eta} = \mathbf{u}|_{\Gamma^\eta(t)},$$

$$\rho_s h \partial_{tt} \boldsymbol{\eta} + \mathcal{L}_e \boldsymbol{\eta} = J \boldsymbol{\sigma} \boldsymbol{n}|_{\Gamma^\eta(t)},$$

where $\boldsymbol{n}$ is the unit outward normal to $\Omega^\eta(t)$, and J denotes the Jacobian of the transformation from Eulerian to Lagrangian coordinates (see (2.2.10)).

In our case, with only radial displacement of the lateral wall different from zero, we have:

$$(\partial_t \eta(t, z), 0) = \boldsymbol{u}(t, z, R + \eta(t, z)), \tag{2.2.8}$$

$$\rho_s h \partial_t^2 \eta + \mathcal{L}_e \eta = -J \, (\boldsymbol{\sigma} \boldsymbol{n})|_{(t, z, R + \eta(t, z))} \cdot \mathbf{e}_r, \tag{2.2.9}$$

where $\mathcal{L}_e$ is given by (2.2.4), and $J = \sqrt{1 + \left(\frac{\partial \eta}{\partial z}\right)^2}$ is the Jacobian of the transformation from the Eulerian to the Lagrangian coordinates. The Jacobian is obtained from the balance of forces in integral form, which has to hold for every

Fig. 2.2 A sketch of the reference boundary Γ and current boundary $\Gamma^\eta(t)$

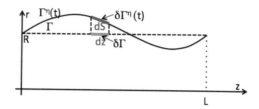

Borel subset $B \subset \Gamma$, and the "corresponding" subset $B(t) \subset \Gamma^\eta(t)$:

$$\int_B \left(\rho_s h \partial_t^2 \eta + \mathcal{L}_e \eta \right) dz = \int_{B(t)} (\boldsymbol{\sigma n})|_{\Gamma^\eta(t)} \cdot \mathbf{e}_r dS$$

$$= \int_B (\boldsymbol{\sigma n})(t, z, R + \eta(t, z)) \cdot \mathbf{e}_r \, J dz,$$

where dS is the surface measure along $\Gamma^\eta(t)$, with

$$dS = \sqrt{1 + \left(\frac{\partial \eta}{\partial z} \right)^2} \, dz = J \, dz. \tag{2.2.10}$$

See Fig. 2.2.

Condition (2.2.8) is the kinematic, and condition (2.2.9) is the dynamic coupling condition. Condition (2.2.9) assumes zero traction from the external environment. Generally, the right hand side of (2.2.9) is given by the **jump** in the traction forces across the interface.

Equations (2.2.2), (2.2.3), (2.2.5), (2.2.6), (2.2.7), (2.2.8), and (2.2.9) define a nonlinear moving-boundary problem for the unknown functions $\mathbf{u}$ and η. The problem is supplemented with initial conditions:

$$\mathbf{u}(0, .) = \mathbf{u}_0, \ \eta(0, .) = \eta_0, \ \partial_t \eta(0, .) = v_0. \tag{2.2.11}$$

We will be assuming that the initial data satisfy the following compatibility conditions:

$$\mathbf{u}_0(z, R + \eta_0(z)) = v_0(z)\mathbf{e}_r, \quad z \in (\Gamma),$$
$$\eta_0(0) = \eta_0(L) = v_0(0) = v_0(L) = 0, \tag{2.2.12}$$
$$R + \eta_0(z) > 0, \quad z \in [0, L].$$

In summary, we study the following nonlinear moving-boundary problem:

Problem 2.1 Find $\mathbf{u} = (u_z(t, z, r), u_r(t, z, r))$, $p(t, z, r)$, and $\eta(t, z)$ such that

$$\left. \begin{array}{c} \rho_f \left(\partial_t \mathbf{u} + (\mathbf{u} \cdot \nabla) \mathbf{u} \right) = \nabla \cdot \boldsymbol{\sigma} \\ \nabla \cdot \mathbf{u} = 0 \end{array} \right\} \ \text{in } \Omega^\eta(t), \ t \in (0, T), \tag{2.2.13}$$

$$\left.\begin{aligned}
\mathbf{u}|_{\Gamma^\eta(t)} &= \partial_t \eta \mathbf{e}_r, \\
\rho_s h \partial_t^2 \eta + C_0 \eta - C_1 \partial_z^2 \eta + C_2 \partial_z^4 \eta &= -J\boldsymbol{\sigma}\mathbf{n}|_{\Gamma^\eta(t)} \cdot \mathbf{e}_r,
\end{aligned}\right\} \quad \text{on } (0, T) \times \Gamma,$$

$$\tag{2.2.14}$$

$$\left.\begin{aligned}
u_r &= 0, \\
\partial_r u_z &= 0,
\end{aligned}\right\} \quad \text{on } (0, T) \times \Gamma_b, \tag{2.2.15}$$

$$\left.\begin{aligned}
p + \frac{\rho_f}{2}|u|^2 &= P_{in/out}(t), \\
u_r &= 0,
\end{aligned}\right\} \quad \text{on } (0, T) \times \Gamma_{in/out}, \tag{2.2.16}$$

$$\eta(t, 0) = \partial_z \eta(t, 0) = \eta(t, L) = \partial_z \eta(t, L) = 0 \quad \text{on } (0, T)$$

$$\left.\begin{aligned}
\mathbf{u}(0, .) &= \mathbf{u}_0, \\
\eta(0, .) &= \eta_0, \\
\partial_t \eta(0, .) &= v_0.
\end{aligned}\right\} \quad \text{at } t = 0. \tag{2.2.17}$$

2.2.3 The Energy of the Coupled Problem

We first show that the formulation of the problem is "reasonable" in the sense that the total energy of the coupled problem is bounded by a constant that depends only on the data. More precisely, we will show that the kinetic energy of the fluid and of the structure, and the elastic energy of the structure, plus fluid dissipation, are all bounded by the L^2-norms of the inlet and outlet data, which depend on time. Here, without loss of generality, we are assuming zero initial data.

Namely, formally we show that the coupled problem (2.2.13)–(2.2.17) satisfies the following energy inequality:

$$\frac{d}{dt}E(t) + D(t) \leq C(P_{in}(t), P_{out}(t)), \tag{2.2.18}$$

where $E(t)$ denotes the sum of the kinetic energy of the fluid and of the structure, and the elastic energy of the Koiter shell:

$$E(t) = \frac{\rho_f}{2}\|\mathbf{u}\|_{L^2(\Omega^\eta(t))}^2 + \frac{\rho_s h}{2}\|\partial_t \eta\|_{L^2(\Gamma)}^2 + \langle \mathcal{L}_e \eta, \eta \rangle, \tag{2.2.19}$$

where

$$\langle \mathcal{L}_e \eta, \eta \rangle := \frac{1}{2}\left(C_0\|\eta\|_{L^2(\Gamma)}^2 + C_1\|\partial_z \eta\|_{L^2(\Gamma)}^2 + C_2\|\partial_z^2 \eta\|_{L^2(\Gamma)}^2\right)$$

for the Koiter shell. The term $D(t)$ captures dissipation due to fluid viscosity:

$$D(t) = \mu \|D(\mathbf{u})\|^2_{L^2(\Omega^\eta(t))}, \tag{2.2.20}$$

and $C(P_{in}(t), P_{out}(t)))$ is a constant which depends only on the inlet and outlet pressure data, and on time $T > 0$.

To show that (2.2.18) holds, we first multiply Eq. (2.2.5) by $\mathbf{u}$, integrate over $\Omega^\eta(t)$, and formally integrate by parts to obtain:

$$\int_{\Omega^\eta(t)} \rho_f \left(\partial_t \mathbf{u} \cdot \mathbf{u} + (\mathbf{u} \cdot \nabla) \mathbf{u} \cdot \mathbf{u} \right) + 2\mu \int_{\Omega^\eta(t)} |D\mathbf{u}|^2 - \int_{\partial\Omega^\eta(t)} (-p\mathbf{I} + 2\mu D(\mathbf{u})) \mathbf{n}(t) \cdot \mathbf{u} = 0.$$

To deal with the inertia term we first recall that $\Omega^\eta(t)$ is moving in time and that the velocity of the lateral boundary is given by $\mathbf{u}|_{\Gamma^\eta(t)}$. The transport theorem, applied to the first term on the left hand-side of the above equation, then gives:

$$\int_{\Omega^\eta(t)} \partial_t \mathbf{u} \cdot \mathbf{u} = \frac{1}{2} \frac{d}{dt} \int_{\Omega^\eta(t)} |\mathbf{u}|^2 - \frac{1}{2} \int_{\Gamma^\eta(t)} |\mathbf{u}|^2 \mathbf{u} \cdot \mathbf{n}(t).$$

The second term on the left hand side can be rewritten by using integration by parts, and the divergence-free condition, to obtain:

$$\int_{\Omega^\eta(t)} (\mathbf{u} \cdot \nabla) \mathbf{u} \cdot \mathbf{u} = \frac{1}{2} \int_{\partial\Omega^\eta(t)} |\mathbf{u}|^2 \mathbf{u} \cdot \mathbf{n}(t) = \frac{1}{2} \left(\int_{\Gamma^\eta(t)} |\mathbf{u}|^2 \mathbf{u} \cdot \mathbf{n}(t) \right.$$

$$\left. - \int_{\Gamma_{in}} |\mathbf{u}|^2 u_z + \int_{\Gamma_{out}} |\mathbf{u}|^2 u_z. \right)$$

These two terms added together give

$$\int_{\Omega^\eta(t)} \partial_t \mathbf{u} \cdot \mathbf{u} + \int_{\Omega^\eta(t)} (\mathbf{u} \cdot \nabla) \mathbf{u} \cdot \mathbf{u} = \frac{1}{2} \frac{d}{dt} \int_{\Omega^\eta(t)} |\mathbf{u}|^2 - \frac{1}{2} \int_{\Gamma_{in}} |\mathbf{u}|^2 u_z + \frac{1}{2} \int_{\Gamma_{out}} |\mathbf{u}|^2 u_z. \tag{2.2.21}$$

To deal with the boundary integral over $\partial\Omega^\eta(t)$, we first notice that on $\Gamma_{in/out}$ the boundary condition (2.2.6) implies $u_r = 0$. Combined with the divergence-free condition we obtain $\partial_z u_z = -\partial_r u_r = 0$. Now, using the fact that the normal to $\Gamma_{in/out}$ is $\mathbf{n} = (\mp 1, 0)$ we get:

$$\int_{\Gamma_{in/out}} (-p\mathbf{I} + 2\mu D(\mathbf{u})) \mathbf{n} \cdot \mathbf{u} = \int_{\Gamma_{in}} P_{in} u_z - \int_{\Gamma_{out}} P_{out} u_z. \tag{2.2.22}$$

In a similar way, using the symmetry boundary conditions (2.2.7), we get:

$$\int_{\Gamma_b} (-p\mathbf{I} + 2\mu\mathbf{D}(\mathbf{u}))\mathbf{n} \cdot \mathbf{u} = 0.$$

What is left is to calculate the remaining boundary integral over $\Gamma^\eta(t)$. For this purpose, we consider the Koiter shell equation (2.2.2), multiplied by $\partial_t \eta$, and integrated by parts to obtain

$$\int_0^L f \partial_t \eta = \frac{\rho_s h}{2} \frac{d}{dt} \|\partial_t \eta\|_{L^2(\Gamma)}^2 + \frac{1}{2} \frac{d}{dt} \left(C_0 \|\eta\|_{L^2(\Gamma)}^2 + C_1 \|\partial_z \eta\|_{L^2(\Gamma)}^2 + C_2 \|\partial_z^2 \eta\|_{L^2(\Gamma)}^2 \right) \tag{2.2.23}$$

By enforcing the coupling conditions (2.2.9) we obtain

$$-\int_{\Gamma^\eta(t)} \sigma \mathbf{n}(t) \cdot \mathbf{u} = -\int_0^L J\sigma \mathbf{n} \cdot \mathbf{u} = \int_0^L f \partial_t \eta. \tag{2.2.24}$$

Finally, by combining (2.2.24) with (2.2.23), and by adding the remaining contributions to the energy of the FSI problem calculated in Eqs. (2.2.21) and (2.2.22), one obtains the following energy equality:

$$\frac{\rho_f}{2} \frac{d}{dt} \int_{\Omega^\eta(t)} |\mathbf{u}|^2 + \frac{\rho_s h}{2} \frac{d}{dt} \|\partial_t \eta\|_{L^2(\Gamma)}^2 + 2\mu \int_{\Omega^\eta(t)} |\mathbf{Du}|^2$$

$$+\frac{1}{2} \frac{d}{dt} \left(C_0 \|\eta\|_{L^2(\Gamma)}^2 + C_1 \|\partial_z \eta\|_{L^2(\Gamma)}^2 + C_2 \|\partial_z^2 \eta\|_{L^2(\Gamma)}^2 \right) \tag{2.2.25}$$

$$= \int_{\Gamma_{out}} P_{out} u_z - \int_{\Gamma_{in}} P_{in} u_z$$

By using the trace inequality [1] and Korn's inequality one can estimate:

$$|P_{in/out}(t)| \int_{\Sigma_{in/out}} u_z| \leq C|P_{in/out}| \|\mathbf{u}\|_{H^1(\Omega^\eta(t))} \leq \frac{C}{2\epsilon} |P_{in/out}|^2 + \frac{\epsilon C}{2} \|\mathbf{D}(\mathbf{u})\|_{L^2(\Omega^\eta(t))}^2.$$

By choosing ϵ such that $\frac{\epsilon C}{2} \leq \mu$ we get the energy inequality (2.2.18).

Remark 2.1 Notice that the constant in the trace inequality depends on the fluid domain, which in our case depends on η. To get an energy estimate in which the constant is independent of η, one can use Gronwall's inequality, and obtain the result above in which the constant C depends on time T.

2.2.4 The FSI Problem Defined on a Fixed Domain

To prove the existence of weak solutions of the coupled FSI problem we need to first address the difficulty related to the motion of the fluid domain. One of the difficulties associated with the motion of the fluid domain is the interpretation of the time derivative of $\mathbf{u}$ when we discretize the problem in time, since the finite difference approximation of the time derivative $(\mathbf{u}^{n+1} - \mathbf{u}^n)/\Delta t$ contains the functions $\mathbf{u}^{n+1}$ and $\mathbf{u}^n$ which are defined on two different domains, one corresponding to the time $t = (n+1)\Delta t$, and the other corresponding to the time $t = n\Delta t$. A way to deal with this issue is to map the family of fluid domains $\Omega^\eta(t)$ onto a fixed (reference) domain and work on the fixed domain. This, in turn, introduces additional (geometric) nonlinearities in the equations. Moreover, the resulting problem no longer has a divergence free fluid velocity given in terms of the divergence operator defined in physical coordinates. This introduces some difficulties in the proof, as we shall see later. This is why in a recent approach to proving existence of weak solution to a nonlinear FSI problem involving mesh-supported shells [18], the authors take a "hybrid" approach, which has been used in numerical schemes for 40 years. The approach is based on temporarily mapping $\mathbf{u}^{n+1}$ and $\mathbf{u}^n$ onto the Lagrangian (fixed, reference) domain, calculating the time derivative there, and then mapping the result to the "current" Eulerian fluid domain Ω^n, and studying the problem on the physical domain Ω^n.

In the present lecture notes we will map the entire problem onto the fixed domain Ω, and show how to deal with the difficulties associated with the divergence free condition, and additional nonlinearities associated with the fluid domain motion.

To map the moving domains $\Omega^\eta(t)$ onto a fixed domain Ω we introduce a family of mappings $\mathcal{A}^\eta$, also known as the Arbitrary Lagrangian Eulerian (ALE) mappings, parameterized by η:

$$\mathcal{A}^\eta(t) : \Omega \to \Omega^\eta(t), \quad \mathcal{A}^\eta(t)(\tilde{z}, \tilde{r}) := \begin{pmatrix} \tilde{z} \\ (R + \eta(t, \tilde{z}))\tilde{r} \end{pmatrix}, \quad (\tilde{z}, \tilde{r}) \in \Omega, \tag{2.2.26}$$

where $(\tilde{z}, \tilde{r})$ denote the coordinates in the reference domain $\Omega = \Gamma \times (0, R)$. Mapping $\mathcal{A}^\eta(t)$ is a bijection (assuming $R + \eta(z, t) > 0$), and its Jacobian is given by

$$\mathcal{J}_{\mathcal{A}^\eta(t)} = |\det \nabla \mathcal{A}^\eta(t)| = |R + \eta(t, \tilde{z})|. \tag{2.2.27}$$

Composite functions with the ALE mapping will be denoted by

$$\mathbf{u}^\eta(t, .) = \mathbf{u}(t, .) \circ \mathcal{A}^\eta(t) \quad \text{and} \quad p^\eta(t, .) = p(t, .) \circ \mathcal{A}^\eta(t). \tag{2.2.28}$$

The derivatives of composite functions satisfy:

$$\partial_t \mathbf{u} = \partial_t \mathbf{u}^\eta - (\mathbf{w}^\eta \cdot \nabla^\eta)\mathbf{u}^\eta, \quad \nabla \mathbf{u} = \nabla^\eta \mathbf{u}^\eta,$$

where the ALE domain velocity, $\mathbf{w}^\eta$, and the transformed gradient, ∇^η, are given by:

$$\mathbf{w}^\eta = \partial_t \eta \tilde{r} \mathbf{e}_r, \quad \nabla^\eta = \begin{pmatrix} \partial_{\tilde{z}} - \tilde{r} \dfrac{\partial_z \eta}{R + \eta} \partial_{\tilde{r}} \\ \dfrac{1}{R + \eta} \partial_{\tilde{r}} \end{pmatrix}. \tag{2.2.29}$$

Note that

$$\nabla^\eta \mathbf{v} = \nabla \mathbf{v} (\nabla \mathcal{A}^\eta)^{-1}. \tag{2.2.30}$$

The following notation will also be useful:

$$\sigma^\eta = -p^\eta \mathbf{I} + 2\mu \mathbf{D}^\eta(\mathbf{u}^\eta), \quad \mathbf{D}^\eta(\mathbf{u}^\eta) = \frac{1}{2}(\nabla^\eta \mathbf{u}^\eta + (\nabla^\eta)^\tau \mathbf{u}^\eta).$$

Remark 2.2 Since our problem is in 2D and allowing only radial displacement to be different from zero, the ALE mapping was easy to construct explicitly. This is, in general, not the case, and various mappings based on elliptic extensions of the boundary data to the fluid domain, may be designed to deal with this issue. For example, often times in numerical solvers the ALE mapping is defined by the harmonic extension of the boundary data onto the fluid domain, i.e., as a solution to the following elliptic problem:

$$\Delta \mathcal{A}^\eta(t) = 0 \text{ on } \Omega,$$
$$\mathcal{A}^\eta(t) = \eta(t) \text{ on } \Gamma,$$
$$\mathcal{A}^\eta(t) = 0 \text{ on } \partial\Omega \setminus \Gamma,$$

calculated at every time step $t = t_n$. Special care needs to be used when designing ALE mappings so that the resulting semi-discretized scheme satisfies the so called (approximate) **geometric conservation law** [36], which is associated with the stability of the scheme.

Problem in ALE Framework: Fixed Domain Thus, the FSI problem (2.2.13)–(2.2.17) in ALE framework, defined on the reference domain Ω, is given by the following: Find $\mathbf{u}(t, \tilde{z}, \tilde{r})$, $p(t, \tilde{z}, \tilde{r})$ and $\eta(t, \tilde{z})$ such that

$$\left. \begin{array}{r} \rho_f \big(\partial_t \mathbf{u} + ((\mathbf{u} - \mathbf{w}^\eta) \cdot \nabla^\eta)\mathbf{u} \big) = \nabla^\eta \cdot \sigma^\eta, \\ \nabla^\eta \cdot \mathbf{u} = 0, \end{array} \right\} \text{ in } (0, T) \times \Omega, \tag{2.2.31}$$

$$\left. \begin{array}{c} u_r = 0, \\ \partial_r u_z = 0 \end{array} \right\} \text{ on } (0, T) \times \Gamma_b, \qquad (2.2.32)$$

$$\left. \begin{array}{c} p + \frac{\rho_f}{2}|u|^2 = P_{in/out}(t), \\ u_r = 0, \end{array} \right\} \text{ on } (0, T) \times \Gamma_{in/out}, \qquad (2.2.33)$$

$$\left. \begin{array}{c} \mathbf{u} = \partial_t \eta \mathbf{e}_r, \\ \rho_s h \partial_{tt} \eta + C_0 \eta - C_1 \partial_z^2 \eta + C_2 \partial_z^4 \eta = -J \sigma \mathbf{n} \cdot \mathbf{e}_r \end{array} \right\} \text{ on } (0, T) \times \Gamma, \qquad (2.2.34)$$

$$\mathbf{u}(0, .) = \mathbf{u}_0, \eta(0, .) = \eta_0, v(0, .) = v_0, \quad \text{at} \quad t = 0. \qquad (2.2.35)$$

Here, we have dropped the superscript η in $\mathbf{u}^\eta$ for easier reading.

Remark 2.3 Since the ALE mapping depends on displacement η, the regularity of the ALE mapping depends on the regularity of η. As we shall see later, to prove the existence of a weak solution to the FSI problem, we will have to assume a certain regularity of the ALE mapping, which will be justified *a posteriori* at the end of the proof, by showing that the resulting solution has the displacement η with sufficient regularity to justify the assumption on the regularity of the ALE mapping.

2.3 Definition of Weak Solutions

We present two weak formulations: one for the problem defined on the family of moving domains $\Omega^\eta(t)$, and one for the problem defined on the fixed, reference domain Ω. Both will be used in the existence proof, presented below. The energy estimate presented above in (2.2.18) motivates the weak solution spaces, as we show next.

2.3.1 Moving Domain Formulation

2.3.1.1 Notation and Function Spaces

To define weak solutions of our moving-boundary problem, the following notation will be useful. First, we introduce the bilinear form associated with the elastic energy of the (clamped) Koiter shell:

$$a_S(\eta, \psi) := \langle \mathcal{L}_e \eta, \psi \rangle = \int_\Gamma \left(C_0 \eta \psi + C_1 \partial_z \eta \partial_z \psi + C_2 \partial_z^2 \eta \partial_z^2 \psi \right), \qquad (2.3.1)$$

and the linear functional which associates the inlet and outlet dynamic pressure boundary data to a test function $\mathbf{v} = (v_z, v_r)$ in the following way:

$$\langle F(t), \mathbf{v} \rangle_{\Gamma_{in/out}} = P_{in}(t) \int_{\Gamma_{in}} v_z - P_{out}(t) \int_{\Gamma_{out}} v_z.$$

To define a weak solution to problem (2.2.13)–(2.2.17) we introduce the following function spaces:

$$\mathcal{V}_F^\eta(t) = \{\mathbf{u} = (u_z, u_r) \in H^1(\Omega^\eta(t))^2 : \nabla \cdot \mathbf{u} = 0,$$
$$u_z = 0 \text{ on } \Gamma^\eta(t), \ u_r = 0 \text{ on } \Omega^\eta(t) \setminus \Gamma^\eta(t)\}, \tag{2.3.2}$$

which is the functions space associated with the fluid velocity, and

$$\mathcal{V}_S = H_0^2(\Gamma), \tag{2.3.3}$$

which is the function space of weak solutions for the Koiter shell. Here $H^1(\Omega^\eta(t))$ is defined by:

$$H^1(\Omega^\eta(t)) = \{\mathbf{u} : \Omega^\eta(t) \to \mathbb{R}^2 \mid \mathbf{u} = \tilde{\mathbf{u}} \circ (\mathcal{A}^\eta)^{-1}, \tilde{\mathbf{u}} \in (H^1(\Omega))^2\}.$$

Motivated by the energy inequality we also define the corresponding evolution spaces for the fluid and structure sub-problems, respectively:

$$\mathcal{W}_F^\eta(0, T) = L^\infty(0, T; (L^2(\Omega^\eta(t))^2)) \cap L^2(0, T; \mathcal{V}_F^\eta(t)) \tag{2.3.4}$$

$$\mathcal{W}_S(0, T) = W^{1,\infty}(0, T; L^2(\Gamma)) \cap H^1(0, T; \mathcal{V}_S). \tag{2.3.5}$$

Using these spaces we can define the solution space for the **coupled fluid-structure interaction problem**, which also includes the kinematic coupling condition:

$$\mathcal{W}(0, T) = \{(\mathbf{u}, \eta) \in \cup_{\eta \in \mathcal{W}_S(0,T)} \mathcal{W}_F^\eta(0, T) \times \{\eta\} : \mathbf{u}(t, z, R + \eta(t, z)) = \partial_t \eta \mathbf{e}_r\}. \tag{2.3.6}$$

The corresponding test spaces for each η will be denoted by

$$Q^\eta(0, T) = \{(\mathbf{q}, \psi) \in C_c^1([0, T); \mathcal{V}_F^\eta \times \mathcal{V}_S) : \mathbf{q}(t, z, R + \eta(t, z)) = \psi(t, z)\mathbf{e}_r\}. \tag{2.3.7}$$

2.3.1.2   Weak Solution: Moving Domain Formulation

We are now in a position to define weak solutions of our moving-boundary problem, defined on the moving domain $\Omega^\eta(t)$.

Definition 2.1 We say that $(\mathbf{u}, \eta) \in \mathcal{W}(0, T)$ is a weak solution of problem (2.2.13)–(2.2.17) if for every $(\mathbf{q}, \psi) \in Q^\eta(0, T)$ the following equality holds:

$$-\rho_f \int_0^T \left(\int_{\Omega^\eta(t)} \mathbf{u} \cdot \partial_t \mathbf{q} \, dx + \frac{1}{2} \int_{\Omega^\eta(t)} (\mathbf{u} \cdot \nabla)\mathbf{u} \cdot \mathbf{q} \, dx - \frac{1}{2} \int_{\Omega^\eta(t)} (\mathbf{u} \cdot \nabla)\mathbf{q} \cdot \mathbf{u} \, dx \right) dt$$

$$+ 2\mu \int_0^T \int_{\Omega^\eta(t)} \mathbf{D}(\mathbf{u}) : \mathbf{D}(\mathbf{q}) dx dt$$

$$- \frac{\rho_f}{2} \int_0^T \int_\Gamma (\partial_t \eta)^2 \psi \, dz dt - \rho_s h \int_0^T \int_\Gamma \partial_t \eta \partial_t \psi \, dz dt + \int_0^T a_S(\eta, \psi) \, dt$$

$$= \int_0^T \langle F(t), \mathbf{q} \rangle_{\Gamma_{in/out}} dt + \rho_f \int_{\Omega_{\eta 0}} \mathbf{u}_0 \cdot \mathbf{q}(0) dx + \rho_s h \int_\Gamma v_0 \psi(0) dz.$$

$$(2.3.8)$$

Equation (2.3.8) is a consequence of integration by parts, and the following equalities which hold for smooth functions:

$$\int_{\Omega^\eta(t)} (\mathbf{u} \cdot \nabla)\mathbf{u} \cdot \mathbf{q} = \frac{1}{2} \int_{\Omega^\eta(t)} (\mathbf{u} \cdot \nabla)\mathbf{u} \cdot \mathbf{q} - \frac{1}{2} \int_{\Omega^\eta(t)} (\mathbf{u} \cdot \nabla)\mathbf{q} \cdot \mathbf{u}$$

$$+ \frac{1}{2} \int_\Gamma (\partial_t \eta)^2 \psi \pm \frac{1}{2} \int_{\Gamma_{out/in}} |u_r|^2 v_r,$$

$$\int_0^T \int_{\Omega^\eta(t)} \partial_t \mathbf{u} \cdot \mathbf{q} = -\int_0^T \int_{\Omega^\eta(t)} \mathbf{u} \cdot \partial_t \mathbf{q} - \int_{\Omega_{\eta 0}} \mathbf{u}_0 \cdot \mathbf{q}(0) - \int_0^T \int_\Gamma (\partial_t \eta)^2 \psi.$$

Remark 2.4 Notice how due to the motion of the fluid domain, we get an extra integral over Γ coming from fluid advection, and accounting for the fluid domain boundary motion via the trace of the fluid velocity, which is equal to the velocity of the domain boundary due to the no-slip condition. This is an important term, which in the derivation of the energy estimate (2.2.18) cancels out the "bad", cubic term, coming from the application of the transport theorem to the kinetic energy term. Therefore, the presence of the nonlinear fluid advection term in the Navier–Stokes equations is crucial in obtaining the energy estimate (2.2.18) for fluid-structure interaction problems on moving domains.

2.3.2 Fixed Domain Formulation

2.3.2.1 Notation and Function Spaces

To define weak solutions on the fixed domain Ω we use the ALE mappings $\mathcal{A}^\eta(t)$ defined in (2.2.26) to map the problem onto Ω via the inverse of $\mathcal{A}^\eta(t)$, and work with the transformed functions, defined in (2.2.28). The first thing to notice is that the transformed fluid velocity $\mathbf{u}^\eta$ is not divergence-free anymore, i.e., $\nabla \cdot \mathbf{u}^\eta \neq 0$. Rather, it satisfies the transformed divergence-free condition $\nabla^\eta \cdot \mathbf{u}^\eta = 0$. Therefore, we need to redefine the function spaces for the fluid velocity by introducing

$$\widetilde{\mathcal{V}}_F^\eta = \{\mathbf{u}^\eta = (u_z^\eta, u_r^\eta) \in (H^1(\Omega))^2 : \nabla^\eta \cdot \mathbf{u}^\eta = 0, \ u_z^\eta = 0 \text{ on } \Gamma, \ u_r^\eta = 0 \text{ on } \Omega \setminus \Gamma\}.$$

The function spaces $\widetilde{\mathcal{W}}_F^\eta(0, T)$ and $\widetilde{\mathcal{W}}(0, T)$ are defined the same way as before, but with $\widetilde{\mathcal{V}}_F^\eta$ instead $\mathcal{V}_F^\eta(t)$. More precisely:

$$\widetilde{\mathcal{W}}_F^\eta(0, T) = L^\infty(0, T; (L^2(\Omega))^2 \cap L^2(0, T; \widetilde{\mathcal{V}}_F^\eta(t)), \tag{2.3.9}$$

$$\widetilde{\mathcal{W}}(0, T) = \{(\mathbf{u}^\eta, \eta) \in \cup_{\eta \in \mathcal{W}_S(0,T)} \widetilde{\mathcal{W}}_F^\eta(0, T) \times \{\eta\} : \mathbf{u}^\eta(t, z, R) = \partial_t \eta(t, z) \mathbf{e}_r\}. \tag{2.3.10}$$

The corresponding test spaces are defined by

$$\widetilde{\mathcal{Q}}^\eta(0, T) = \{(\mathbf{q}, \psi) \in C_c^1([0, T); \widetilde{\mathcal{V}}_F^\eta \times \mathcal{V}_S) : \mathbf{q}(t, z, R) = \psi(t, z) \mathbf{e}_r\}. \tag{2.3.11}$$

The weak formulation on the fixed domain is defined next.

2.3.2.2 Weak Solution: Fixed Domain Formulation

To obtain the weak formulation on the fixed domain we begin by transforming the first integral on the left hand-side of (2.3.8) in Definition 2.1, describing the fluid kinetic energy. Namely, by recalling the definition of the Jacobian of the ALE transformation (2.2.27), which is equal to $R + \eta$, we formally calculate:

$$-\int_{\Omega^\eta(t)} \mathbf{u}^\eta \cdot \partial_t \mathbf{q} d\mathbf{x} = -\int_\Omega \mathbf{u}^\eta \cdot (\partial_t \mathbf{q} - (\mathbf{w}^\eta \cdot \nabla^\eta) \mathbf{q})(R + \eta) d\tilde{\mathbf{x}} = -\int_\Omega (R + \eta) \mathbf{u}^\eta \cdot \partial_t \mathbf{q} d\tilde{\mathbf{x}}$$

$$+ \frac{1}{2} \int_\Omega (R + \eta)(\mathbf{w}^\eta \cdot \nabla^\eta) \mathbf{q} \cdot \mathbf{u}^\eta d\tilde{\mathbf{x}} + \frac{1}{2} \int_\Omega (R + \eta)(\mathbf{w}^\eta \cdot \nabla^\eta) \mathbf{q} \cdot \mathbf{u}^\eta d\tilde{\mathbf{x}}.$$

In the last integral on the right hand-side we use the definition of $\mathbf{w}^\eta$ and of ∇^η, given in (2.2.29), to obtain

$$\int_\Omega (R + \eta)(\mathbf{w}^\eta \cdot \nabla^\eta) \mathbf{q} \cdot \mathbf{u}^\eta d\tilde{\mathbf{x}} = \int_\Omega \partial_t \eta \, \tilde{r} \, \partial_{\tilde{r}} \mathbf{q} \cdot \mathbf{u}^\eta d\tilde{\mathbf{x}}.$$

Using integration by parts with respect to r, and keeping in mind that η does not depend on r, we obtain

$$-\int_{\Omega^\eta(t)} \mathbf{u}^\eta \cdot \partial_t \mathbf{q} = -\int_\Omega (R + \eta)\mathbf{u}^\eta \cdot \partial_t \mathbf{q} + \frac{1}{2}\int_\Omega (R + \eta)(\mathbf{w}^\eta \cdot \nabla^\eta)\mathbf{q} \cdot \mathbf{u}^\eta$$

$$-\frac{1}{2}\int_\Omega (R + \eta)(\mathbf{w}^\eta \cdot \nabla^\eta)\mathbf{u}^\eta \cdot \mathbf{q} - \frac{1}{2}\int_\Omega \partial_t \eta \mathbf{u}^\eta \cdot \mathbf{q} + \frac{1}{2}\int_0^L (\partial_t \eta)^2 \psi,$$

By using this calculation and the definition of weak solutions Definition 2.1, the following weak formulation on fixed domain follows.

Definition 2.2 We say that $(\mathbf{u}, \eta) \in \widetilde{\mathcal{W}}(0, T)$ is a weak solution of problem (2.2.31)–(2.2.35) defined on the reference domain Ω, if for every $(\mathbf{q}, \psi) \in \widetilde{\mathcal{Q}^\eta}(0, T)$ the following equality holds:

$$-\rho_f \int_0^T \int_\Omega (R + \eta)\mathbf{u} \cdot \partial_t \mathbf{q} d\tilde{x} dt + \frac{\rho_f}{2}\int_0^T \int_\Omega (R + \eta)\big(((\mathbf{u} - \mathbf{w}^\eta) \cdot \nabla^\eta)\mathbf{u} \cdot \mathbf{q}$$

$$-((\mathbf{u} - \mathbf{w}^\eta) \cdot \nabla^\eta)\mathbf{q} \cdot \mathbf{u})\big) d\tilde{x} dt - \frac{\rho_f}{2}\int_0^T \int_\Omega (\partial_t \eta)\mathbf{u} \cdot \mathbf{q} d\tilde{x} dt$$

$$+2\mu \int_0^T \int_\Omega (R + \eta)\mathbf{D}^\eta(\mathbf{u}) : \mathbf{D}^\eta(\mathbf{q}) d\tilde{x} dt - \rho_s h \int_0^T \int_\Gamma \partial_t \eta \partial_t \psi dz dt + \int_0^T a_S(\eta, \psi) dt$$

$$= R \int_0^T \Big(P_{in}(t) \int_0^R (q_z)_{|z=0} dr - P_{out}(t) \int_0^R (q_z)_{|z=L} dr \Big) dt$$

$$+ \rho_f \int_{\Omega^{\eta_0}} \mathbf{u}_0 \cdot \mathbf{q}(0) d\tilde{x} + \rho_s h \int_\Gamma v_0 \psi(0) dz.$$

$$(2.3.12)$$

2.4 The Lie Operator Splitting

To find a weak solution to the coupled FSI problem we first construct approximate solutions by semi-discretizing the problem, written as a first-order system, in time. This is sometimes called the Rothe's method. While discretizing the problem in time, we also separate the fluid from structure sub-problems using a Lie operator splitting strategy [43]. Existence of weak solutions is obtained by showing that there exist sub-sequences of those approximate solutions, which converge to a weak solution as the time-discretization step goes to zero.

2.4.1 General Definition of the Splitting Scheme

The Lie splitting, also known as the Marchuk–Yanenko splitting scheme, can be summarized as follows. Let $N \in \mathbb{N}$, $\Delta t = T/N$ and $t_n = n\Delta t$. Consider the following initial-value problem:

$$\frac{d\phi}{dt} + A\phi = 0 \quad \text{in } (0, T), \quad \phi(0) = \phi_0,$$

where A is an operator defined on a Hilbert space, and A can be written as $A = A_1 + A_2$. Set $\phi^0 = \phi_0$, and, for $n = 0, \ldots, N - 1$ and $i = 1, 2$, compute $\phi^{n+\frac{i}{2}}$ by solving

$$\left.\begin{array}{c} \dfrac{d}{dt}\phi_i + A_i\phi_i = 0 \\[2mm] \phi_i(t_n) = \phi^{n+\frac{i-1}{2}} \end{array}\right\} \quad \text{in } (t_n, t_{n+1}),$$

and then set $\phi^{n+\frac{i}{2}} = \phi_i(t_{n+1})$, for $i = 1, 2$. It can be shown that this method is at least $\sqrt{\Delta t}$-accurate in time, see e.g., [43].

We apply this splitting to out coupled FSI problem (2.2.31)–(2.2.35) by first rewriting the problem as a first-order system in time. Then, operators A_1 and A_2 are defined to correspond to a structure sub-problem and a fluid sub-problem, respectively. The solution of the structure sub-problem will be used as the initial data for the fluid sub-problem, and the solution of the fluid sub-problem will be used as the initial data for the structure sub-problem *at the next time step*. No sub-iterations between the fluid and structure sub-problems are needed. In this splitting, the fluid and structure communicate only via the initial data. The trick is to define the structure and fluid sub-problems in such a way that the resulting scheme converges to a weak solution of the coupled FSI. This is highly non-trivial, and will require some physical intuition to motivate the splitting, as we explain below.

2.4.2   The Coupled Problem in First-Order Form

To apply the Lie splitting strategy we must first write the coupled FSI problem as a first-order system in time. For this purpose we introduce a new variable: the structure velocity

$$v = \partial_t \eta = \partial_t \eta \mathbf{e}_r.$$

Now, the coupled FSI problem in first-order form reads:

$$\rho_f \partial_t \mathbf{u} = \rho_f (\mathbf{u} \cdot \nabla)\mathbf{u} + \nabla \cdot \boldsymbol{\sigma}, \text{ with contraint } \nabla \cdot \mathbf{u} = 0 \quad \text{in } \Omega^\eta(t), \ t \in (0, T),$$

$$\left.\begin{array}{l} \partial_t \eta = \mathbf{u}|_{\Gamma^\eta(t)} \\[1mm] \rho_s h \partial_t v = -\mathcal{L}_e \eta + \sigma n|_{\Gamma^\eta(t)} \\[1mm] \partial_t \eta = v \end{array}\right\} \quad \text{on } \Gamma, \ t \in (0, T). \tag{2.4.1}$$

Problem in First-Order Form: Fixed Domain The initial-boundary value problem defined on the fixed domain, in first-order form reads: Find $\mathbf{u}(t, \tilde{z}, \tilde{r})$, $p(t, \tilde{z}, \tilde{r})$, $\eta(t, \tilde{z})$, and $v(t, \tilde{z})$ such that

$$\left. \begin{aligned} \rho_f \left(\partial_t \mathbf{u} + ((\mathbf{u} - \mathbf{w}^\eta) \cdot \nabla^\eta) \mathbf{u} \right) &= \nabla^\eta \cdot \sigma^\eta, \\ \nabla^\eta \cdot \mathbf{u} &= 0, \end{aligned} \right\} \text{ in } (0, T) \times \Omega, \tag{2.4.2}$$

$$\left. \begin{aligned} u_r &= 0, \\ \partial_r u_z &= 0 \end{aligned} \right\} \text{ on } (0, T) \times \Gamma_b, \tag{2.4.3}$$

$$\left. \begin{aligned} p + \tfrac{\rho_f}{2} |u|^2 &= P_{in/out}(t), \\ u_r &= 0, \end{aligned} \right\} \text{ on } (0, T) \times \Gamma_{in/out}, \tag{2.4.4}$$

$$\left. \begin{aligned} \mathbf{u} &= v e_r, \\ \partial_t \eta &= v, \\ \rho_s h \partial_t v + C_0 \eta - C_1 \partial_z^2 \eta + C_2 \partial_z^4 \eta &= -J \sigma \mathbf{n} \cdot e_r \end{aligned} \right\} \text{ on } (0, T) \times \Gamma, \tag{2.4.5}$$

$$\mathbf{u}(0, .) = \mathbf{u}_0, \eta(0, .) = \eta_0, v(0, .) = v_0, \quad \text{at} \quad t = 0. \tag{2.4.6}$$

2.4.3 The Splitting and the Added Mass Effect

There are many different ways to split the coupled problem (2.4.2)–(2.4.6). Perhaps the "obvious" way would be to keep the kinematic coupling condition, i.e., the no-slip condition, together with the fluid sub-problem and solve the Navier–Stokes equations on the 'current" domain with the Dirichlet condition for the fluid velocity describing no-slip. Then calculate the trace of the normal stress $\sigma \mathbf{n}$ on the boundary Γ from the just calculated fluid velocity, and use it to load the structure. In this splitting, the kinematic coupling condition would be associated with the fluid sub-problem, and the entire dynamic coupling condition would be considered as the structure sub-problem. This splitting is known in numerical methods as the Dirichlet–Neumann scheme, since it uses Dirichlet data for the fluid sub-problem, and then takes the solution of the fluid sub-problem to calculate the trace of the normal stress on the boundary, which is the Neumann data, to load the structure. The operator which maps the boundary trace of the solution of the Dirichlet problem to the corresponding Neumann boundary data is called the Dirichlet–Neumann operator. The splitting described above is the Dirichlet–Neumann splitting.

Although this splitting may seem reasonable, it is now well-known that it suffers from stability issues in numerical "loosely-coupled schemes", namely, in partitioned schemes in which there are no sub-iterations between the fluid and structure sub-problems. This is especially the case when the fluid and structure have comparable densities and are particularly sensitive to the so called *added mass effect* [20]. In the Dirichlet–Neumann splitting, the structure velocity *from the previous time step*

n is used as the no-slip condition in the fluid sub-problem at the *next time step*, $n + 1$, giving the fluid normal stress at $n + 1$ which, in turn, depends on the structure velocity from the time step n. When this normal stress is used to load the structure at time $n + 1$, the resulting scheme for the structure displacement is *explicit*: it consists of the structure inertia term and the elastic energy terms expressed at time $n + 1$, and the loading term coming from the fluid, which depends on the structure velocity at time n via the Dirichlet–Neumann operator. This loading term incorporates the fluid inertia (in which the fluid velocity at the interface is substituted by the structure velocity at time n), and acts as an extra mass term (added mass) in the structure subproblem. This added mass term accounts for the fact that the structure immersed in the fluid needs to displace the surrounding fluid around it as it moves, thereby increasing its inertia due to the fluid added mass. It was shown in [20] that the resulting scheme is unconditionally unstable for problems in which the fluid and structure densities are comparable. The Dirichlet–Neumann approach introduces poor approximation of the energy exchange between the fluid and structure in the coupled problem, and without sub-iterations to correct for this problem, the resulting scheme is unstable. By keeping the fluid and structure inertia tied together implicitly, this problem is avoided. This is why we designed a Lie splitting in which the fluid and structure inertia are tied implicitly. More precisely, we split the structure equation into two parts: the part that incorporates fluid information, and the part that incorporates only structure information:

$$
\left.
\begin{array}{c}
\rho_s h \partial_t v + \underbrace{\mathcal{L}_e \eta}_{Structure} = \underbrace{-J\sigma \mathbf{n}|_{\Gamma^\eta(t)} \cdot \mathbf{e}_r}_{Fluid} \\[2em]
\partial_t \eta = \underbrace{v}_{Structure} \\[2em]
\partial_t v \cdot \mathbf{e}_r = \underbrace{\mathbf{u}|_{\Gamma^\eta(t)}}_{Fluid}
\end{array}
\right\} \quad \text{on } (0, T) \times \Gamma.
$$

Here we used the notation $\mathbf{u}|_{\Gamma^\eta(t)}$ to denote the composite function with the ALE mapping $\mathcal{A}^\eta(t)$:

$$
\mathbf{u}|_{\Gamma^\eta(t)} := \mathbf{u} \circ \mathcal{A}^\eta(t) = \mathbf{u}(t, R + \eta(z, t), z), \quad z \in \Gamma.
$$

The part that incorporates fluid information with be used as the boundary condition in the fluid sub-problem, and the part that incorporates structure information will define the structure sub-problem. As dictated by the Lie operator splitting, the first-order derivatives on the left hand-side will contribute to both the fluid and structure sub-problems. More precisely, the following is the splitting of the coupling conditions that will enter definitions of the operators A_1 (structure) and A_2 (fluid) in the Lie splitting:

The Structure Sub-Problem (Operator A_1)

$$\left. \begin{array}{l} \rho_s h \partial_t v + \mathcal{L}_e \eta = 0 \\ \partial_t \eta = v \end{array} \right\} \quad \text{which implies} \quad \partial_{tt}\eta + \mathcal{L}_e \eta = 0 \text{ on } (0, T) \times \Gamma,$$

which will be solved with the initial data for the structure velocity $\partial_t \eta = v$ given by the trace of the fluid velocity $\mathbf{u}|_{\Gamma^\eta(t)} \cdot \mathbf{e}_r$ on Γ obtained from the previous time step.

The Fluid Sub-Problem (Operator A_2)
The fluid sub-problem consists of solving the Navier–Stokes equations with the boundary condition on the moving boundary given by

$$\left\{ \begin{array}{l} \rho_s h \partial_t v = -J \boldsymbol{\sigma} \mathbf{n}|_{\Gamma^\eta(t)} \cdot \mathbf{e}_r \\ \mathbf{u}|_{\Gamma^\eta(t)} = v \mathbf{e}_r \end{array} \right\} \quad \text{or} \quad \left\{ \begin{array}{l} \rho_s h \partial_t u_r|_{\Gamma^\eta(t)} = -J \boldsymbol{\sigma} \mathbf{n}|_{\Gamma^\eta(t)} \cdot \mathbf{e}_r \\ u_l|_{\Gamma^\eta(t)} = 0 \end{array} \right\}.$$

Notice that instead of the Dirichlet boundary condition for the Navier–Stokes equations, we now have a Robin-type boundary condition. The boundary condition incorporates the structure inertia term $\rho_s h \partial_t u_r|_{\Gamma^\eta(t)}$ (which is equal to $\rho_s h \partial_t v$), which will be evaluated at time $n + 1$. This will lead to an implicit scheme in the structure subproblem where the fluid loading via the fluid normal stress on Γ, will have the information from the time step $n + 1$, and will be implicitly tied to the rest of the structure problem.

We make this operator splitting strategy precise in the next section.

2.4.4 The Splitting in Semi-Discretized Form

We define the Lie operator splitting for the problem defined on the fixed domain Ω, and use the Backward Euler scheme to approximate the time derivatives in the fluid and structure sub-problems.

Let Δt denote the time step and $N \in \mathbb{N}$ the number of time sub-intervals, so that

$$(0, T) = \cup_{n=0}^{N-1}(t^n, t^{n+1}), \quad t^n = n\Delta t, \ n = 0, \ldots, N - 1.$$

For every subdivision containing $N \in \mathbb{N}$ sub-intervals, we recursively define the vector of unknown approximate solutions

$$\mathbf{X}_N^{n+\frac{i}{2}} = \begin{pmatrix} \mathbf{u}_N^{n+\frac{i}{2}} \\ v_N^{n+\frac{i}{2}} \\ \eta_N^{n+\frac{i}{2}} \end{pmatrix}, n = 0, 1, \ldots, N - 1, \ i = 1, 2, \tag{2.4.7}$$

where $i = 1, 2$ denotes the solution of sub-problem A1 or A2, respectively. The initial condition will be denoted by

$$\mathbf{X}^0 = \begin{pmatrix} \mathbf{u}_0 \\ v_0 \\ \eta_0 \end{pmatrix}.$$

As hinted earlier, the semi-discretization and the splitting of the problem will be performed in such a way that the discrete version of the energy inequality (2.2.18) is preserved at every time step. This is a crucial ingredient for the existence proof. For this purpose, we define the semi-discrete versions of the kinetic and elastic energy, defined in (2.2.19), and of dissipation, defined in (2.2.20), as follows:

$$
E_N^{n+\frac{i}{2}} = \frac{1}{2}\left(\rho_f \int_\Omega (R + \eta_N^n)|\mathbf{u}_N^{n+\frac{i}{2}}|^2 + \rho_s h \|v_N^{n+\frac{i}{2}}\|_{L^2(\Gamma)}^2 \right.
$$
$$
\left. + C_0\|\eta_N^{n+\frac{i}{2}}\|_{L^2(\Gamma)}^2 + C_1\|\partial_z \eta_N^{n+\frac{i}{2}}\|_{L^2(\Gamma)}^2 + C_2\|\partial_z^2 \eta_N^{n+\frac{i}{2}}\|_{L^2(\Gamma)}^2 \right),
$$
(2.4.8)

$$
D_N^{n+1} = \Delta t \mu \int_\Omega (R + \eta_N^n)|D^{\eta_N^n}(\mathbf{u}_N^{n+1})|^2, \quad n = 0, \dots, N-1, \ i = 0, 1.
$$
(2.4.9)

Throughout the rest of this section, we fix the time step Δt, i.e., we keep $N \in \mathbb{N}$ fixed, and study the semi-discretized sub-problems defined by the Lie splitting. To simplify notation, we will omit the subscript N and write $(\mathbf{u}^{n+\frac{i}{2}}, v^{n+\frac{i}{2}}, \eta^{n+\frac{i}{2}})$ instead of $(\mathbf{u}_N^{n+\frac{i}{2}}, v_N^{n+\frac{i}{2}}, \eta_N^{n+\frac{i}{2}})$.

2.4.4.1 Operator A_1: The Structure Sub-Problem

In this step $\mathbf{u}$ does not change, and so

$$\mathbf{u}^{n+\frac{1}{2}} = \mathbf{u}^n.$$

We define $(v^{n+\frac{1}{2}}, \eta^{n+\frac{1}{2}}) \in H_0^2(\Gamma) \times H_0^2(\Gamma)$ as a solution of the following problem, written in weak form:

$$
\int_\Gamma \frac{\eta^{n+\frac{1}{2}} - \eta^n}{\Delta t} \phi \, dz = \int_\Gamma v^{n+\frac{1}{2}} \phi \, dz, \quad \phi \in L^2(\Gamma),
$$
(2.4.10)

$$
\rho_s h \int_\Gamma \frac{v^{n+\frac{1}{2}} - v^n}{\Delta t} \psi \, dz + a_S(\eta^{n+\frac{1}{2}}, \psi) = 0, \quad \psi \in H_0^2(\Gamma).
$$

The first equation is a weak form of the semi-discretized kinematic coupling condition, while the second equation corresponds to a weak form of the semi-discretized elastodynamics equation with zero forcing on the right hand-side.

Proposition 2.1 *For each fixed $\Delta t > 0$, problem (2.4.10) has a unique solution $(v^{n+\frac{1}{2}}, \eta^{n+\frac{1}{2}}) \in H_0^2(\Gamma) \times H_0^2(\Gamma)$.*

Proof The proof is a direct consequence of the Lax–Milgram lemma applied to the weak form

$$\int_0^L \eta^{n+\frac{1}{2}} \psi dz + (\Delta t)^2 a_S(\eta^{n+\frac{1}{2}}, \psi) = \int_0^L \left(\Delta t v^n + \eta^n \right) \psi dz, \quad \psi \in H_0^2(\Gamma),$$

which is obtained after eliminating $v^{n+\frac{1}{2}}$ in the second equation by using the kinematic coupling condition given by the first equation.

Proposition 2.2 *For each fixed $\Delta t > 0$, solution of problem (2.4.10) satisfies the following discrete energy equality:*

$$\begin{aligned}
E_N^{n+\frac{1}{2}} &+ \frac{1}{2}\left(\rho_s h\|v^{n+\frac{1}{2}} - v^n\|^2 + C_0\|\eta^{n+\frac{1}{2}} - \eta^n\|^2 \right. \\
&\left. + C_1\|\partial_z(\eta^{n+\frac{1}{2}} - \eta^n)\|^2 + C_2\|\partial_z^2(\eta^{n+\frac{1}{2}} - \eta^n)\|^2\right) = E_N^n,
\end{aligned} \tag{2.4.11}$$

where the kinetic energy E_N^n is defined in (2.4.8).

Proof From the first equation in (2.4.10) we immediately get

$$v^{n+\frac{1}{2}} = \frac{\eta^{n+\frac{1}{2}} - \eta^n}{\Delta t} \in H_0^2(\Gamma).$$

Therefore we can take $v^{n+\frac{1}{2}}$ as a test function in the second equation in (2.4.10). We replace the test function ψ by $v^{n+\frac{1}{2}}$ in the first term on the left hand-side, and replace ψ by $(\eta^{n+\frac{1}{2}} - \eta^n)/\Delta t$ in the bilinear form a_S. We then use the algebraic identity $(a - b) \cdot a = \frac{1}{2}(|a|^2 + |a - b|^2 - |b|^2)$ to deal with the terms $(v^{n+1/2} - v^n)v^{n+1/2}$ and $(\eta^{n+1/2} - \eta^n)\eta^{n+1/2}$. After multiplying the entire equation by Δt, the second equation in (2.4.10) can be written as:

$$\rho_s h(\|v^{n+\frac{1}{2}}\|^2 + \|v^{n+\frac{1}{2}} - v^n\|^2) + a_S(\eta^{n+\frac{1}{2}}, \eta^{n+\frac{1}{2}}) + a_S(\eta^{n+\frac{1}{2}} - \eta^n, \eta^{n+\frac{1}{2}} - \eta^n)$$

$$= \rho_s h\|v^n\|^2 + a_S(\eta^n, \eta^n).$$

We then recall that $\mathbf{u}^{n+\frac{1}{2}} = \mathbf{u}^n$ in this sub-problem, and so we can add $\rho_f \int_\Omega (1 + \eta^n)\mathbf{u}^{n+1/2}$ on the left hand-side, and $\rho_f \int_\Omega (1 + \eta^n)\mathbf{u}^n$ on the right hand-side of the equation, to obtain exactly the energy equality (2.4.11).

2.4.4.2 Operator A_2: The Fluid Sub-Problem

In this step η does not change, and so

$$\eta^{n+1} = \eta^{n+\frac{1}{2}}.$$

To define the weak formulation for the fluid sub-problem, we introduce the following function spaces:

$$\begin{aligned}
(\widetilde{\mathcal{V}}_F^{\eta})_N^{n+1} &= \{(\mathbf{u}^{n+1}, v^{n+1}) \in \widetilde{\mathcal{V}}_F^{\eta^n} \times L^2(\Gamma) \ : \ \mathbf{u}_{|\Gamma}^{n+1} = v^{n+1}\mathbf{e}_r\}, \\
(\widetilde{Q}^{\eta})_N^{n+1} &= \{(\mathbf{q}, \psi) \in \widetilde{\mathcal{V}}_F^{\eta^n} \times H_0^2(\Gamma) \ : \ \mathbf{q}_{|\Gamma} = \psi\mathbf{e}_r\}.
\end{aligned} \tag{2.4.12}$$

The weak formulation for the fluid sub-problem is then given by: find $(\mathbf{u}^{n+1}, v^{n+1}) \in (\widetilde{\mathcal{V}}_F^{\eta})_N^{n+1}$ such that $\forall (\mathbf{q}, \psi) \in (\widetilde{Q}^{\eta})_N^{n+1}$ the following holds:

$$\rho_f \int_\Omega (R + \eta^n) \left(\frac{\mathbf{u}^{n+1} - \mathbf{u}^{n+\frac{1}{2}}}{\Delta t} \cdot \mathbf{q} + \frac{1}{2} \left[(\mathbf{u}^n - v^{n+\frac{1}{2}}r\mathbf{e}_r) \cdot \nabla^{\eta^n} \right] \mathbf{u}^{n+1} \cdot \mathbf{q} \right.$$

$$\left. -\frac{1}{2} \left[(\mathbf{u}^n - v^{n+\frac{1}{2}}r\mathbf{e}_r) \cdot \nabla^{\eta^n} \right] \mathbf{q} \cdot \mathbf{u}^{n+1} \right) d\widetilde{\mathbf{x}} + \frac{\rho_f}{2} \int_\Omega v^{n+\frac{1}{2}} \mathbf{u}^{n+1} \cdot \mathbf{q} d\widetilde{\mathbf{x}}$$

$$+2\mu \int_\Omega (R + \eta^n) \mathbf{D}^{\eta^n}(\mathbf{u}) : \mathbf{D}^{\eta^n}(\mathbf{q}) d\widetilde{\mathbf{x}} + \rho_s h \int_\Gamma \frac{v^{n+1} - v^{n+\frac{1}{2}}}{\Delta t} \psi dz$$

$$= R \left(P_{in}^n \int_0^R (q_z)_{|z=0} dr - P_{out}^n \int_0^R (q_z)_{|z=L} dr \right),$$

$$\text{with } \nabla^{\eta^n} \cdot \mathbf{u}^{n+1} = 0,$$

$$\tag{2.4.13}$$

where $P_{in/out}^n = \frac{1}{\Delta t} \int_{n\Delta t}^{(n+1)\Delta t} P_{in/out}(t) dt.$

Proposition 2.3 *Let $\Delta t > 0$, and assume that η^n are such that $R + \eta^n \geq R_{\min} > 0, n = 0, \ldots, N$. Then, the fluid sub-problem defined by (2.4.13) has a unique weak solution $(\mathbf{u}^{n+1}, v^{n+1}) \in (\widetilde{Q}^{\eta})_N^{n+1}$.*

Proof The proof is again a consequence of the Lax–Milgram lemma. More precisely, for $(\mathbf{u}, v), (\mathbf{q}, \psi) \in (\widetilde{Q}^{\eta})_N^{n+1}$ define the bilinear form associated with problem (2.4.13):

$$
\begin{aligned}
\mathcal{B}((\mathbf{u}, v), (\mathbf{q}, \psi)) := {} & \rho_f \int_\Omega (R + \eta^n) \left(\mathbf{u} \cdot \mathbf{q} + \frac{\Delta t}{2} \left[(\mathbf{u}^n - v^{n+\frac{1}{2}} r \mathbf{e}_r) \cdot \nabla^{\eta^n} \right] \mathbf{u} \cdot \mathbf{q} \right. \\
& \left. - \frac{\Delta t}{2} \left[(\mathbf{u}^n - v^{n+\frac{1}{2}} r \mathbf{e}_r) \cdot \nabla^{\eta^n} \right] \mathbf{q} \cdot \mathbf{u} \right) \\
& + \Delta t \frac{\rho_f}{2} \int_\Omega v^{n+\frac{1}{2}} \mathbf{u} \cdot \mathbf{q} + \Delta t 2 \mu \int_\Omega (R + \eta^n) \mathbf{D}^{\eta^n}(\mathbf{u}) : \mathbf{D}^{\eta^n}(\mathbf{q}) \\
& + \rho_s h \int_\Gamma v \psi.
\end{aligned}
$$

We need to prove that this bilinear form $\mathcal{B}$ is coercive and continuous on $(\widetilde{Q}^n)_N^{n+1}$. To see that $\mathcal{B}$ is coercive, we write

$$
\mathcal{B}((\mathbf{u}, v), (\mathbf{u}, v)) = \rho_f \int_\Omega (R + \eta^n + \frac{\Delta t}{2} v^{n+\frac{1}{2}}) |\mathbf{u}|^2 + \rho_s h \int_\Gamma v^2 + \Delta t 2 \mu \int_\Omega (R + \eta^n) |\mathbf{D}^{\eta^n}(\mathbf{u})|^2.
$$

Coercivity follows immediately after recalling that η^n are such that $R + \eta^n \geq R_{\min} > 0$, which implies that $R + \eta^n + \frac{\Delta t}{2} v^{n+\frac{1}{2}} = R + \frac{1}{2}(\eta^n + \eta^{n+\frac{1}{2}}) \geq R_{\min} > 0$.

Before we prove continuity of $\mathcal{B}$, notice that from (2.2.29) we have:

$$
\| \nabla^{\eta^n} \mathbf{u} \|_{L^2(\Omega)} \leq C \| \eta^n \|_{H^2(\Gamma)} \| \mathbf{u} \|_{H^1(\Omega)}.
$$

Therefore, by applying the generalized Hölder inequality and the continuity of the embedding of H^1 into L^4, we obtain

$$
\begin{aligned}
\mathcal{B}((\mathbf{u}, v), (\mathbf{q}, \psi)) \leq {} & C \left(\rho_f \| \mathbf{u} \|_{L^2(\Omega)} \| \mathbf{q} \|_{L^2(\Omega)} + \rho_s h \| v \|_{L^2(\Gamma)} \| \psi \|_{L^2(\Gamma)} \right. \\
& + \Delta t \| \eta^n \|_{H^2(\Gamma)} (\| \mathbf{u}^n \|_{H^1(\Omega)} + \| v^{n+\frac{1}{2}} \|_{H^1(\Gamma)}) \| \mathbf{u} \|_{H^1(\Omega)} \| \mathbf{q} \|_{H^1(\Omega)} \\
& \left. + \Delta t \mu \| \eta^n \|_{H^2(\Gamma)}^2 \| \mathbf{u} \|_{H^1(\Omega)} \| \mathbf{q} \|_{H^1(\Omega)} \right).
\end{aligned}
$$

This shows that $\mathcal{B}$ is continuous. The Lax–Milgram lemma now implies the existence of a unique solution $(\mathbf{u}^{n+1}, v^{n+1})$ of problem (2.4.13).

Proposition 2.4 *For each fixed $\Delta t > 0$, solution of problem (2.4.13) satisfies the following discrete energy inequality:*

$$
\begin{aligned}
E_N^{n+1} + \frac{\rho_f}{2} \int_\Omega (R + \eta^n) |\mathbf{u}^{n+1} - \mathbf{u}^n|^2 + \frac{\rho_s h}{2} \| v^{n+1} - v^{n+\frac{1}{2}} \|_{L^2(\Gamma)}^2 & \\
+ D_N^{n+1} \leq E_N^{n+\frac{1}{2}} + C \Delta t ((P_{in}^n)^2 + (P_{out}^n)^2), &
\end{aligned} \tag{2.4.14}
$$

where the kinetic energy E_N^n and dissipation D_N^n are defined in (2.4.8) and (2.4.9), and the constant C depends only on the parameters in the problem, and not on Δt (or N).

Proof We begin by focusing on the weak formulation (2.4.13) in which we replace the test functions $\mathbf{q}$ by $\mathbf{u}^{n+1}$ and ψ by v^{n+1}. We multiply the resulting equation by Δt, and notice that the first term on the right hand-side is given by

$$\frac{\rho_f}{2}\int_\Omega (R+\eta^n)|\mathbf{u}^{n+1}|^2.$$

This is the term that contributes to the discrete kinetic energy at the time step $n+1$, but it does not have the correct form, since the discrete kinetic energy at $n+1$ is given in terms of the structure location at $n+1$, and not at n, namely, the discrete kinetic energy at $n+1$ is:

$$\frac{\rho_f}{2}\int_\Omega (R+\eta^{n+1})|\mathbf{u}^{n+1}|^2.$$

To get around this difficulty it is crucial that the advection term is present in the fluid sub-problem. The advection term is responsible for the presence of the integral

$$\frac{\rho_f}{2}\int_\Omega \Delta t v^{n+\frac{1}{2}}|\mathbf{u}^{n+1}|^2$$

which can be re-written by noticing that $\Delta t v^{n+\frac{1}{2}} := (\eta^{n+1/2}-\eta^n)$ which is equal to $(\eta^{n+1}-\eta^n)$ since, in this sub-problem $\eta^{n+1}=\eta^{n+1/2}$. This implies

$$\frac{\rho_f}{2}\left(\int_\Omega (R+\eta^n)|\mathbf{u}^{n+1}|^2 + \Delta t v^{n+\frac{1}{2}}|\mathbf{u}^{n+1}|^2\right) = \frac{\rho_f}{2}\int_\Omega (R+\eta^{n+1})|\mathbf{u}^{n+1}|^2.$$

Thus, these two terms combined provide the discrete kinetic energy at the time step $n+1$. It is interesting to notice how the nonlinearity of the coupling at the deformed boundary requires the presence of nonlinear advection in order for the discrete kinetic energy of the fluid sub-problem to be bounded and decreasing in time.

To complete the proof one simply uses the algebraic identity $(a-b)\cdot a = \frac{1}{2}(|a|^2 + |a-b|^2 - |b|^2)$ in the same way as in the proof of Proposition 2.2.

2.4.5 Uniform Energy Estimates for the Coupled Semi-Discretized Problem

To get the uniform energy estimates for the coupled, semi-discretized problem, we first define the weak form of the problem. For this purpose, we will be using the function spaces defined in (2.4.12). The weak form of the *semi-discretized coupled problem* on a *fixed domain* is given by the following: find $(\mathbf{u}^{n+1}, v^{n+1}) \in (\tilde{\mathcal{V}}_F^n)_N^{n+1}$

and $(v^{n+\frac{1}{2}}, \eta^{n+\frac{1}{2}}) \in H_0^2(\Gamma) \times H_0^2(\Gamma)$ such that

$$
\begin{aligned}
\rho_f \int_\Omega (R + \eta_N^n) \frac{\mathbf{u}_N^{n+1} - \mathbf{u}_N^n}{\Delta t} \cdot \mathbf{q} &+ \frac{\varrho_f}{2} \int_\Omega (R + \eta_N^n) \Big(\Big[(\mathbf{u}_N^n - v_N^{n+1/2} r \mathbf{e}_r) \cdot \nabla \eta_N^n \Big] \mathbf{u}_N^{n+1} \cdot \mathbf{q} \\
&- \Big[(\mathbf{u}_N^n - v_N^{n+1/2} r \mathbf{e}_r) \cdot \nabla \eta_N^n \Big] \mathbf{q} \cdot \mathbf{u}_N^{n+1} \Big) \\
&+ \frac{\rho_f}{2} \int_\Omega v_N^{n+1/2} \mathbf{u}_N^{n+1} \cdot \mathbf{q} + 2\mu \int_\Omega (R + \eta_N^n) \mathbf{D}^{\eta_N^n}(\mathbf{u}_N^{n+1}) : \mathbf{D}^{\eta_N^n}(\mathbf{q}) \\
&+ \rho_s h \int_\Gamma \frac{v_N^{n+1} - v_N^n}{\Delta t} \psi + a_e(\eta_N^{n+1}, \psi) = F^n(\mathbf{q}), \\
\int_\Gamma \frac{\eta_N^{n+1} - \eta_N^n}{\Delta t} \psi &= \int_\Gamma v_N^{n+1/2} \psi, \quad \forall (\mathbf{q}, \psi) \in Q_N^n, \\
&\text{with } \nabla \eta_N^n \cdot \mathbf{u}_N^{n+1} = 0,
\end{aligned}
\tag{2.4.15}
$$

and

$$
F^n(\mathbf{q}) := R \Big(P_{in}^n \int_{\Gamma_{in}} (q_z)|_{\Gamma_{in}} - P_{out}^n \int_{\Gamma_{out}} (q_z)|_{\Gamma_{out}} \Big),
$$

where $P_{in/out}^n$ denote the piecewise constant approximations of $P_{in/out}(t)$.

Recall that $\mathbf{u}_N^{n+1}$ is defined via the ALE mapping associated with the "previous" (known) domain $\Omega^{\eta_N^n}$.

The uniform energy estimates for the coupled problem are obtained by combining the energy estimates for the fluid and structure sub-problems, discussed above.

Lemma 2.1 (Uniform Energy Estimates) *Let $\Delta t > 0$ and $N = T/\Delta t > 0$. Furthermore, let $E_N^{n+\frac{1}{2}}$, E_N^{n+1}, and D_N^j be the kinetic energy and dissipation given by* (2.4.8) *and* (2.4.9), *respectively.*

There exists a constant $C > 0$ independent of Δt (and N), which depends only on the parameters in the problem, on the kinetic energy of the initial data E_0, and on the energy norm of the inlet and outlet data $\| P_{in/out} \|_{L^2(0,T)}^2$, such that the following estimates hold:

1. *Kinetic and elastic energy estimate: $E_N^{n+\frac{1}{2}} \le C$, $E_N^{n+1} \le C$, $n = 0, \ldots, N - 1$;*
2. *Viscous fluid dissipation estimate: $\sum_{j=1}^N D_N^j \le C$;*
3. *Numerical dissipation estimates:*

$$
\sum_{n=0}^{N-1} \Big(\int_\Omega (R + \eta^n) |\mathbf{u}^{n+1} - \mathbf{u}^n|^2 + \| v^{n+1} - v^{n+\frac{1}{2}} \|_{L^2(\Gamma)}^2 + \| v^{n+\frac{1}{2}} - v^n \|_{L^2(\Gamma)}^2 \Big) \le C
$$

$$
\sum_{n=0}^{N-1} \Big(C_0 \| \eta^{n+1} - \eta^n \|_{L^2(\Gamma)}^2 + C_1 \| \partial_z(\eta^{n+1} - \eta^n) \|_{L^2(\Gamma)}^2 + C_2 \| \partial_z^2(\eta^{n+1} - \eta^n) \|_{L^2(\Gamma)}^2 \Big) \le C.
$$

In fact, $C = E_0 + \widetilde{C}\left(\|P_{in}\|^2_{L^2(0,T)} + \|P_{out}\|^2_{L^2(0,T)}\right)$, *where* $\widetilde{C}$ *is the constant from* (2.4.14), *which depends only on the parameters in the problem and T.*

Proof We begin by adding the energy estimates (2.4.11) and (2.4.14) to obtain

$$E_N^{n+1} + D_N^{n+1} + \frac{1}{2}\left(\rho_f \int_\Omega (R + \eta^n)|\mathbf{u}^{n+1} - \mathbf{u}^n|^2 + \rho_s h\|v^{n+1} - v^{n+\frac{1}{2}}\|^2_{L^2(\Gamma)} +\right.$$

$$+\rho_s h\|v^{n+\frac{1}{2}} - v^n\|^2_{L^2(\Gamma)} + C_0\|\eta^{n+\frac{1}{2}} - \eta^n\|^2_{L^2(\Gamma)} + C_1\|\partial_z(\eta^{n+\frac{1}{2}} - \eta^n)\|^2_{L^2(\Gamma)} +$$

$$\left.+C_2\|\partial_z^2(\eta^{n+\frac{1}{2}} - \eta^n)\|^2_{L^2(\Gamma)}\right) \le E_N^n + \widetilde{C}\Delta t((P_{in}^n)^2 + (P_{out}^n)^2), \quad n = 0, \ldots, N-1.$$

Then we calculate the sum, on both sides, and cancel the same terms in the kinetic energy that appear on both sides of the inequality to obtain

$$E_N^N + \sum_{n=0}^{N-1} D_N^{n+1} + \frac{1}{2}\sum_{n=0}^{N-1}\left(\rho_f \int_\Omega (R + \eta^n)|\mathbf{u}^{n+1} - \mathbf{u}^n|^2 + \rho_s h\|v^{n+1} - v^{n+\frac{1}{2}}\|^2_{L^2(\Gamma)} +\right.$$

$$+\rho_s h\|v^{n+\frac{1}{2}} - v^n\|^2_{L^2(\Gamma)} + C_0\|\eta^{n+\frac{1}{2}} - \eta^n\|^2_{L^2(\Gamma)} + C_1\|\partial_z(\eta^{n+\frac{1}{2}} - \eta^n)\|^2_{L^2(\Gamma)} +$$

$$\left.+C_2\|\partial_z^2(\eta^{n+\frac{1}{2}} - \eta^n)\|^2_{L^2(\Gamma)}\right) \le E_0 + \widetilde{C}\Delta t \sum_{n=0}^{N-1}((P_{in}^n)^2 + (P_{out}^n)^2).$$

To estimate the term involving the inlet and outlet pressure we recall that on every sub-interval (t^n, t^{n+1}) the pressure data is approximated by a constant which is equal to the average value of the pressure over that time interval. Therefore, we have, after using Hölder's inequality:

$$\Delta t \sum_{n=0}^{N-1}(P_{in}^n)^2 = \Delta t \sum_{n=0}^{N-1}\left(\frac{1}{\Delta t}\int_{n\Delta t}^{(n+1)\Delta t} P_{in}(t)dt\right)^2 \le \|P_{in}\|^2_{L^2(0,T)}.$$

By using the pressure estimate to bound the right hand-side in the above energy estimate, we have obtained all the statements in the Lemma, with the constant C given by $C = E_0 + \widetilde{C}\|P_{in/out}\|^2_{L^2(0,T)}$.

Notice that Statement 1 can be obtained in the same way by summing from 0 to $n - 1$, for each n, instead of from 0 to $N - 1$.

The uniform energy estimates for the coupled problem will be crucial in obtaining weak- and weak*-convergent subsequences of approximate solutions, for which we will show converge to a weak solution by using a generalization of the Aubin–Lions–Simon compactness lemma [67] to problems on moving domains. We will use Lemma 2.1 in the next section to show weak- and weak*-convergence of approximate solutions.

2.5 Weak and Weak* Convergence of Approximate Functions

We pause for a second, and summarize what we have accomplished so far. For a given $\Delta t > 0$ we divided the time interval $(0, T)$ into $N = T/\Delta t$ sub-intervals $(t^n, t^{n+1}), n = 0, \ldots, N - 1$. On each sub-interval (t^n, t^{n+1}) we "solved" the coupled FSI problem by applying the Lie splitting scheme. First we solved for the structure position (Problem A1) and then for the fluid flow (Problem A2). We have just shown that each sub-problem has a unique solution (as a function of $\tilde{x}$), provided that $R + \eta^n \geq R_{\min} > 0, n = 0, \ldots, N$, and that its solution satisfies an energy estimate. When combined, the two energy estimates provide a time-discrete version of the energy estimate (2.2.18), which is obtained in Lemma 2.1. Thus, for each Δt (i.e., for each sub-division N of the time interval $(0, T)$) we have a time-marching, splitting scheme which defines approximations at each t^n, which are functions of the spatial variable. What we would like to ultimately show is that, as $\Delta t \to 0$, a sequence of solutions parameterized by N (or Δt) and defined by those approximations, converges to a weak solution of the problem. Furthermore, we also need to show that $R + \eta^n \geq R_{\min} > 0$ is satisfied for each $n = 0, \ldots, N - 1$. To accomplish these goals, we first define approximate *functions/solutions* which depend not only on $\tilde{x}$ but also on t.

2.5.1 Approximate Functions

We define approximate solutions by using piecewise constant extensions in time of the functions defined for each t^n, so that on each sub-interval $((n - 1)\Delta t, n\Delta t] \subset (0, T)$ we have (see Fig. 2.3):

$$\mathbf{u}_N(t, .) = \mathbf{u}_N^n, \eta_N(t, .) = \eta_N^n, v_N(t, .) = v_N^n, v_N^*(t, .) = v_N^{n-\frac{1}{2}}, t \in ((n - 1)\Delta t, n\Delta t].$$
$$(2.5.1)$$

Notice that functions $v_N^* = v_N^{n-1/2}$ are determined by Step A_1 (the elastodynamics sub-problem), while functions $v_N = v_N^n$ are determined by Step A_2 (the fluid sub-problem). As a consequence, only the functions v_N are equal to the trace of the fluid velocity on Γ, i.e., $\mathbf{u}_N = v_N \mathbf{e}_r$. This is not necessarily the case for the functions v_N^*. However, we will show later that the difference between the two sequences converges to zero in L^2. Using Lemma 2.1 we now show that these sequences are uniformly bounded in the appropriate function spaces.

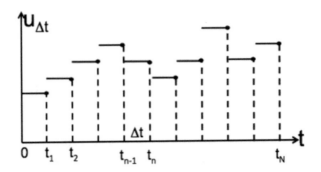

Fig. 2.3 A sketch of u_N

2.5.2 Uniform Boundedness and a Uniform Korn's Estimate

We begin by showing that $(\eta_N)_{N \in \mathbb{N}}$ is uniformly bounded in $L^\infty(0, T; H_0^2(\Gamma))$, and that there exists a $T > 0$ for which $R + \eta_N^n > 0$ holds independently of N and n. This implies, among other things, that our approximate solutions are, indeed, well-defined on a non-zero time interval $(0, T)$.

Proposition 2.5 *Sequence* $(\eta_N)_{N \in \mathbb{N}}$ *is uniformly bounded in*

$$L^\infty(0, T; H_0^2(\Gamma)).$$

Moreover, for T small enough, we have

$$0 < R_{\min} \le R + \eta_N(t, z) \le R_{\max}, \quad \forall N \in \mathbb{N}, z \in (\Gamma), t \in (0, T). \tag{2.5.2}$$

Proof From Lemma 2.1 we have that $E_N^n \le C$, where C is independent of N. This implies

$$\|\eta_N(t)\|_{L^2(\Gamma)}^2, \ \|\partial_z \eta_N(t)\|_{L^2(\Gamma)}^2, \ \|\partial_{zz}^2 \eta_N(t)\|_{L^2(\Gamma)}^2 \le C, \ \forall t \in [0, T].$$

Therefore,

$$\|\eta_N\|_{L^\infty(0, T; H_0^2(\Gamma))} \le C.$$

To show that the radius $R + \eta_N$ is uniformly bounded away from zero for T small enough, we first notice that the above inequality implies

$$\|\eta_N^n - \eta_0\|_{H_0^2(\Gamma)} \le 2C, \ n = 1, \ldots, N, \ N \in \mathbb{N},$$

where we recall that $\eta_N^0 = \eta_0$. Furthermore, we calculate

$$\|\eta_N^n - \eta_0\|_{L^2(\Gamma)} \leq \sum_{i=0}^{n-1} \|\eta_N^{i+1} - \eta_N^i\|_{L^2(\Gamma)} = \Delta t \sum_{i=0}^{n-1} \|v_N^{i+\frac{1}{2}}\|_{L^2(\Gamma)}.$$

From Lemma 2.1 we have that $E_N^{n+\frac{1}{2}} \leq C$, where C is independent of N. Combined with the above inequality, this implies

$$\|\eta_N^n - \eta_0\|_{L^2(\Gamma)} \leq Cn\Delta t \leq CT, \quad n = 1, \ldots, N, \quad N \in \mathbb{N}.$$

Now that we have uniform bounds for $\|\eta_N^n - \eta_0\|_{L^2(\Gamma)}$ and $\|\eta_N^n - \eta_0\|_{H_0^2(\Gamma)}$, we can use the interpolation inequality for Sobolev spaces (see for example [1], Thm. 4.17, p. 79) to get

$$\|\eta_N^n - \eta_0\|_{H^1(\Gamma)} \leq 2C\sqrt{T}, \quad n = 1, \ldots, N, \quad N \in \mathbb{N}.$$

From Lemma 2.1 we see that $C = C(T)$ depends on T through the norms of the inlet and outlet data in such a way that C is an increasing function of T. Therefore by choosing T small, we can make $\|\eta_N^n - \eta_0\|_{H^1(\Gamma)}$ arbitrary small for $n = 1, \ldots, N$, $N \in \mathbb{N}$. Because of the Sobolev embedding of $H^1(\Gamma)$ into $C[0, L]$ we can also make $\|\eta_N^n - \eta_0\|_{C[0,L]}$ arbitrary small. Since the initial data η_0 is such that $R + \eta_0(z) > 0$ (due to the conditions listed in (2.2.12)), we see that for a $T > 0$ small enough, there exist $R_{\min}, R_{\max} > 0$, such that

$$0 < R_{\min} \leq R + \eta_N(t, z) \leq R_{\max}, \quad \forall N \in \mathbb{N}, z \in (\Gamma), t \in (0, T).$$

We will show in the end, see Theorem 2.3, that our existence result holds not only locally in time, i.e., for small $T > 0$, but rather, it can be extended all the way until either $T = \infty$, or until the lateral walls of the channel touch each other.

From this Proposition we see that the L^2-norm $\|f\|_{L^2(\Omega)} = \int f^2 d\tilde{x}$, and the weighted L^2-norm $\|f\|_{L^2(\Omega)} = \int (R + \eta_N) f^2 d\tilde{x}$ are equivalent. More precisely, for every $f \in L^2(\Omega)$, there exist constants $C_1, C_2 > 0$, which depend only on $R_{\min}, R_{\max}$, and not on f or N, such that

$$C_1 \int_\Omega (R + \eta_N) f^2 d\tilde{x} \leq \|f\|_{L^2(\Omega)}^2 \leq C_2 \int_\Omega (R + \eta_N) f^2 d\tilde{x}. \tag{2.5.3}$$

We will be using this property in the next section to prove strong convergence of approximate functions.

Next we show that the sequences of approximate solutions for the fluid and structure velocities are uniformly bounded.

Proposition 2.6 *The following statements hold:*

1. $(v_N)_{n \in \mathbb{N}}$ *is uniformly bounded in* $L^\infty(0, T; L^2(\Gamma)) \cap L^2(0, T; H_0^2(\Gamma))$.
2. $(v_N^*)_{n \in \mathbb{N}}$ *is uniformly bounded in* $L^\infty(0, T; L^2(\Gamma))$.
3. $(\mathbf{u}_N)_{n \in \mathbb{N}}$ *is uniformly bounded in* $L^\infty(0, T; L^2(\Omega)) \cap L^2(0, T; H^1(\Omega))$.

Proof The uniform boundedness of $(v_N)_{N \in \mathbb{N}}$, $(v_N^*)_{N \in \mathbb{N}}$, and the uniform boundedness of $(\mathbf{u}_N)_{N \in \mathbb{N}}$ in $L^\infty(0, T; L^2(\Omega))$ follow directly from Statements 1 and 2 of Lemma 2.1, and from the definition of $(v_N)_{n \in \mathbb{N}}$, $(v_N^*)_{N \in \mathbb{N}}$ and $(\mathbf{u}_N)_{N \in \mathbb{N}}$ as step-functions in t so that

$$\int_0^T \|v_N\|_{L^2(\Gamma)}^2 dt = \sum_{n=0}^{N-1} \|v_N^n\|_{L^2(\Gamma)}^2 \Delta t.$$

To show uniform boundedness of $(\mathbf{u}_N)_{N \in \mathbb{N}}$ in $L^2(0, T; H^1(\Omega))$ we need to explore the boundedness of $(\nabla \mathbf{u}_N)_{N \in \mathbb{N}}$. From Lemma 2.1 we only know that the symmetrized gradient is bounded in the following way:

$$\sum_{n=1}^N \int_\Omega (R + \eta_N^{n-1}) |\mathbf{D}^{\eta_N^{n-1}} (\mathbf{u}_N^n)|^2 d\tilde{\mathbf{x}} \Delta t \leq C. \tag{2.5.4}$$

Remark 2.5 (Korn's Inequality) We cannot immediately apply Korn's inequality since estimate (2.5.4) is given in terms of the transformed symmetrized gradient. Moreover, the Korn's constant depends on the fluid domain, and we need a uniform Korn's constant, independent of the family of moving domains to get a uniform bound of the gradient. Thus, there are some technical difficulties that need to be overcome due to the fact that our problem is defined on a sequence of moving domains, and we would like to obtain a uniform in N bound for the gradient $(\nabla \mathbf{u}_N)_{N \in \mathbb{N}}$.

To get around these difficulties we take the following approach. We first transform the problem back into the original domain $\Omega^{\eta_N^{n-1}}$ on which $\mathbf{u}_N$ is defined, and apply the Korn's inequality in the usual way. However, since the Korn constant depends on the domain, we will need a result which provides a universal Korn constant, independent of the family of domains under consideration.

We first transform the problem onto the physical domain. For each fixed $N \in \mathbb{N}$, and for all $n = 1, \ldots, N$, we map the fluid velocity $\mathbf{u}_N^n$ to the physical domain, which is determined by the location of η_N^{n-1}. We will be using N as a *superscript* to denote the corresponding functions defined on physical domains:

$$\mathbf{u}^{N,n} := \mathbf{u}_N^n \circ (\mathcal{A}^{\eta_N^{n-1}})^{-1}, \ n = 1, \ldots, N, \ N \in \mathbb{N}.$$

By using formula (2.2.29) we get

$$\int_\Omega (1 + \eta_N^{n-1}) |\mathbf{D}^{\eta_N^{n-1}}(\mathbf{u}_N^n)|^2 = \int_{\Omega^{\eta_N^{n-1}}} |\mathbf{D}(\mathbf{u}^{N,n})|^2 = \|\mathbf{D}(\mathbf{u}^{N,n})\|^2_{L^2(\Omega^{\eta_N^{n-1}})}.$$

Now, we want to design a Korn's inequality in such a way that the Korn's constant is independent of N.

There are two sets of references that address this issue. One is the approach from [21, 62, 76], which assumes certain domain regularity. In our case, the conditions of [76] translate into requiring that the family of fluid domains $\{\Omega^{\eta_N^{n-1}}\}_{N \in \mathbb{N}}$ has a uniform Lipschitz constant, which is indeed the case, and the proof using the result from [76] can be found in [62].

Another approach, which is more straight-forward, and does not require the uniform Lischitz property of the family of fluid domains, is an approach based on directly showing that under the following two conditions:

- The fluid velocity is divergence-free;
- The tangential component of the trace of the fluid velocity on the moving boundary is zero;

the L^2-norm of the velocity gradient is equal to 2 times the L^2-norm of the symmetrized gradient of velocity. This is similar to the result by Chambolle et al. in [21], Lemma 6, pg. 377.

Proposition 2.7 (Uniform Korn Equality) *Let* $\mathbf{u} \in \mathcal{V}_F^\eta(t)$ *where* $\mathcal{V}_F^\eta(t)$ *is defined in (2.3.2). Then, the following Korn-type equality holds:*

$$\|\nabla \mathbf{u}\|^2_{L^2(\Omega^\eta)} = 2 \|\mathbf{D}(\mathbf{u})\|^2_{L^2(\Omega^\eta)}. \tag{2.5.5}$$

Notice that the Korn constant (the number 2) is domain independent. The proof of this Korn equality is similar to the proof in Chambolle et al. [21]. However, since our assumptions are a somewhat different from those in [21], we present the proof here.

Proof By writing the symmetrized gradient on the right hand-side of (2.5.5) in terms of the gradient, and by calculating the square of the norms on both sides, one can see that it is sufficient to show that

$$\int_{\Omega^\eta} \nabla \mathbf{u} : \nabla^\tau \mathbf{u} = 0.$$

First, we prove the above equality for smooth functions $\mathbf{u}$ and then the conclusion follows by a density argument. By using integration by parts and $\nabla \cdot \mathbf{u} = 0$ we get

$$\int_{\Omega_F^\eta} \nabla \mathbf{u} : \nabla^\tau \mathbf{u} = -\int_{\Omega_F^\eta} \mathbf{u} \cdot \nabla(\nabla \cdot \mathbf{u}) + \int_{\partial\Omega_F^\eta} (\nabla^\tau \mathbf{u})\mathbf{n} \cdot \mathbf{u} = \int_{\partial\Omega_F^\eta} (\nabla^\tau \mathbf{u})\mathbf{n} \cdot \mathbf{u},$$

where $\mathbf{n} = (-\eta_z, 1)^\tau$. We now show that $(\nabla^\tau \mathbf{u})\mathbf{n} \cdot \mathbf{u} = 0$ on $\partial\Omega_F$. Since $\partial\Omega_F = \Gamma^\eta \cup \Gamma_{in/out} \cup \Gamma_b$ we consider each part of the boundary separately:

1. On Γ^η we have $\mathbf{u} = (0, u_r)$, i.e., we have $u_z(z, R + \eta(z)) = 0$. Since $\mathbf{u}$ is smooth we can differentiate this equality w.r.t. z to get $\partial_z u_z + \partial_r u_z \eta_z = 0$ on Γ, i.e., for $z \in (\Gamma)$. By using $\nabla \cdot \mathbf{u} = 0$, we get: $-\partial_r u_z \eta_z = \partial_z u_z = -\partial_r u_r$. Since $\mathbf{n} = (-\eta_z, 1)^\tau$ we get

$$(\nabla^\tau \mathbf{u})\mathbf{n} \cdot \mathbf{u} = ((\nabla^\tau \mathbf{u})\mathbf{n})_r u_r = (-\partial_r u_z \eta_z + \partial_r u_r)u_r = 0.$$

2. On $\Gamma_{in/out}$ we have $\mathbf{u} = (u_z, 0)$ and $\mathbf{n} = (\pm 1, 0)$. Hence,

$$(\nabla^\tau \mathbf{u})\mathbf{n} \cdot \mathbf{u} = u_z((\nabla^\tau \mathbf{u})\mathbf{n})_z = u_z(\partial_z u_z) = -u_z \partial_r u_r = 0.$$

3. On Γ_b we have $\mathbf{u} = (u_z, 0)$, $\partial_r u_z = 0$ and $\mathbf{n} = (0, -1)$. Hence,

$$(\nabla^\tau \mathbf{u})\mathbf{n} \cdot \mathbf{u} = u_z((\nabla^\tau \mathbf{u})\mathbf{n})_z = u_z(-\partial_z u_r) = 0.$$

This concludes the proof of Korn's equality (2.5.5).

Now, by using (2.5.5) and by mapping everything back to the fixed domain Ω, we obtain the following Korn's equality on Ω:

$$\int_\Omega (1 + \eta_N^{n-1})|\nabla^{\eta_N^{n-1}}(\mathbf{u}_N^n)|^2 = 2\int_\Omega (1 + \eta_N^{n-1})|\mathbf{D}^{\eta_N^{n-1}}(\mathbf{u}_N^n)|^2. \tag{2.5.6}$$

Notice that on the left hand-side we still have the transformed gradient $\nabla_N^{\eta_N^{n-1}}$ and not ∇, and so we employ (2.2.30) to calculate the relationship between the two:

$$\nabla \mathbf{u}_N^n = \left(\nabla^{\eta_N^{n-1}} \mathbf{u}_N^n\right)\left(\nabla \mathscr{A}^{\eta_N^{n-1}}\right), \quad n = 1, \ldots, N, \ N \in \mathbb{N}.$$

Since η_N are bounded in $L^\infty(0, T; H^2(\Gamma))$, the gradient of the ALE mapping is bounded:

$$\|\nabla \mathscr{A}^{\eta_N^{n-1}}\|_{L^\infty(\Omega)} \leq C, \quad n = 1, \ldots, N, \ N \in \mathbb{N}.$$

Using this estimate, and by summing from $n = 1, \ldots, N$, we obtain the following estimate for $\nabla \mathbf{u}_N^n$:

$$\sum_{n=1}^N \|\nabla \mathbf{u}_N^n\|_{L^2(\Omega)}^2 \Delta t \leq C \sum_{n=1}^N \int_\Omega (R + \eta_N^{n-1})|\mathbf{D}^{\eta_N^{n-1}}(\mathbf{u}_N^n)|^2 d\tilde{x} \Delta t,$$

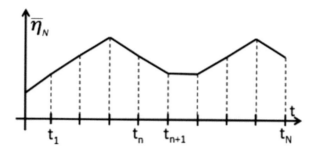

Fig. 2.4 A sketch of $\bar{\eta}_N$

where C is independent of N. This implies that the sequence $(\nabla \mathbf{u}_N)_{N \in \mathbb{N}}$ is uniformly bounded in $L^2((0, T) \times \Omega)$, and so the sequence $(\mathbf{u}_N)_{N \in \mathbb{N}}$ is uniformly bounded in $L^2(0, T; H^1(\Omega))$.

Remark 2.6 (Additional Regularity in Time) To pass to the limit in the weak formulations of approximate solutions, we need additional regularity in time of the structure displacement.

To get higher regularity in time, we introduce a slightly different extrapolation in time of the approximate functions $\eta_N^n, n = 1, \ldots, N$ to the time interval $(0, T)$. Instead of the piece-wise constant extrapolation, we now introduce the *continuous, piecewise linear* extrapolation, which defines a new approximate function $\bar{\eta}_N$ on $(0, T)$. See Fig. 2.4. We note that the final result is independent of the type of extrapolation used.

Definition 2.3 (Piecewise Linear Extrapolation) For each fixed Δt we define $\bar{\eta}_N$ to be the continuous, piece-wise linear function on $(0, T)$, such that on each sub-interval $[(n-1)\Delta t, n\Delta t]$, with $n = 1, \ldots, N$, the linear extrapolation satisfies:

$$\bar{\eta}_N(n\Delta t, .) = \eta_N(n\Delta t, .).$$

We now observe that

$$\partial_t \bar{\eta}_N(t) = \frac{\eta_N^{n+1} - \eta_N^n}{\Delta t} = \frac{\eta_N^{n+1/2} - \eta_N^n}{\Delta t} = v_N^{n+\frac{1}{2}}, \ t \in (n\Delta t, (n+1)\Delta t),$$

and so, since v_N^* was defined in (2.5.1) as a piece-wise constant function defined via $v_N^*(t, \cdot) = v^{n+\frac{1}{2}}$, for $t \in (n\Delta t, (n+1)\Delta t]$, we see that

$$\partial_t \bar{\eta}_N = v_N^* \ a.e. \text{ on } (0, T). \tag{2.5.7}$$

By using Lemma 2.1 (the boundedness of $E_N^{n+\frac{i}{2}}$), we get

$(\bar\eta_N)_{N\in\mathbb{N}}$ is bounded in $L^\infty(0, T; H_0^2(\Gamma)) \cap W^{1,\infty}(0, T; L^2(\Gamma))$.

We have just shown the following

Proposition 2.8 *The sequence $(\bar\eta_N)_{N\in\mathbb{N}}$ is uniformly bounded in $L^\infty(0, T; H_0^2(\Gamma))$ $\cap W^{1,\infty}(0, T; L^2(\Gamma))$.*

We will use this proposition to show precompactness of $(\eta_N)_{N\in\mathbb{N}}$ in $L^\infty(0, T; H^s(\Gamma))$, $0 < s < 2$.

2.5.3   Weakly and Weakly* Convergent Subsequences

From the uniform boundedness of approximate sequences, one obtains weakly or weakly* convergent subsequences (which we denote the same way as the original sequence). More precisely, we have the following result.

Lemma 2.2 *There exist subsequences $(\eta_N)_{N\in\mathbb{N}}$, $(\bar\eta_N)_{N\in\mathbb{N}}$, $(v_N)_{N\in\mathbb{N}}$, $(v_N^*)_{N\in\mathbb{N}}$, and $(\mathbf{u}_N)_{N\in\mathbb{N}}$, and the functions $\eta \in L^\infty(0, T; H_0^2(\Gamma))$, $v \in L^\infty(0, T; L^2(\Gamma)) \cap L^2(0, T; H_0^2(\Gamma))$, $v^* \in L^\infty(0, T; L^2(\Gamma))$, and $\mathbf{u} \in L^\infty(0, T; L^2(\Omega)) \cap L^2(0, T; H^1(\Omega))$, such that*

$$\begin{aligned}
\eta_N &\rightharpoonup \eta \text{ weakly* in } L^\infty(0, T; H_0^2(\Gamma)),\\
\eta_N &\rightharpoonup \eta \text{ weakly* in } W^{1,\infty}(0, T; L^2(\Gamma)),\\
\bar\eta_N &\rightharpoonup \bar\eta \text{ weakly* in } L^\infty(0, T; L^2(\Gamma)),\\
v_N &\rightharpoonup v \text{ weakly* in } L^\infty(0, T; L^2(\Gamma)),\\
v_N^* &\rightharpoonup v^* \text{ weakly* in } L^\infty(0, T; L^2(\Gamma)),\\
\mathbf{u}_N &\rightharpoonup \mathbf{u} \text{ weakly* in } L^\infty(0, T; L^2(\Omega)),\\
\mathbf{u}_N &\rightharpoonup \mathbf{u} \text{ weakly in } L^2(0, T; H^1(\Omega)).
\end{aligned} \tag{2.5.8}$$

Furthermore,

$$\bar\eta = \eta \text{ and } v = v^*. \tag{2.5.9}$$

Proof The statement $\bar\eta = \eta$ is a trivial consequence of the uniqueness of the distributional limit. As a consequence, we get that the weak* limit of η_N is, in fact, in $W^{1,\infty}(0, T; L^2(\Gamma))$.

The only thing left to show is that $v = v^*$. To show this, we multiply the second statement in Lemma 2.1 by Δt, and notice again that $\|v_N\|_{L^2((0,T)\times(\Gamma))}^2 = \Delta t \sum_{n=1}^{N} \|v_N^n\|_{L^2(\Gamma)}^2$. This implies $\|v_N - v_N^*\|_{L^2((0,T)\times(\Gamma))} \le C\sqrt{\Delta t}$, and so in the limit as $\Delta t \to 0$ we get $v = v^*$.

2.6 Strong Convergence of Approximate Functions

To show that sequences/subsequences constructed above converge to a weak solution of our coupled FSI problem, we need to pass to the limit, as $\Delta t \to 0$, or $N \to \infty$, in the weak formulations of the semi-discretized, coupled FSI problems (2.4.15), and show that the limits satisfy the weak formulation of the continuous problem (2.3.12). Unfortunately, due to the presence of nonlinear terms, weak convergence is not sufficient to allow us to conclude that the weak limits satisfy the weak formulation (2.3.12). For this purpose we will show that there exist sub-sequences of the approximations of the fluid and structure velocities, and sub-sequences of the approximate structure displacements, which converge strongly in the appropriate topologies.

   The compactness arguments used to obtain strong convergence for the fluid and structure velocities will rely on the use of a recent generalization of the Aubin–Lions–Simon compactness lemma to problems on moving domains [67], while the compactness arguments for the structure displacement will rely on the Arzelà–Ascoli theorem.

2.6.1 Strong Convergence of Structure Displacements

Recall that, from Proposition 2.8, $(\bar{\eta}_N)_{N\in\mathbb{N}}$ is uniformly bounded in $L^\infty(0, T; H_0^2(\Gamma)) \cap W^{1,\infty}(0, T; L^2(\Gamma))$. Now, by the interpolation estimate, see Theorem 5.2, [1], there exists a constant $C > 0$ independent of N, such that

$$\|\bar{\eta}_N(t+\Delta t) - \bar{\eta}_N(t)\|_{H^{2\alpha}(\Gamma)} \leq C\|\bar{\eta}_N(t+\Delta t) - \bar{\eta}_N(t)\|_{L^2(\Gamma)}^{1-\alpha}\|\bar{\eta}_N(t+\Delta t) - \bar{\eta}_N(t)\|_{H^2(\Gamma)}^{\alpha}.$$

By multiplying and dividing the right hand-side by $(\Delta t)^{1-\alpha}$, and using uniform boundedness of v_N, we get:

$$\|\bar{\eta}_N(t+\Delta t) - \bar{\eta}_N(t)\|_{H^{2\alpha}(\Gamma)} \leq C(\Delta t)^{1-\alpha}, \quad \text{where } 0 < \alpha < 1. \qquad (2.6.1)$$

Therefore, $(\bar{\eta}_N)_{N\in\mathbb{N}}$ is uniformly bounded in $C^{0,1-\alpha}([0, T]; H^{2\alpha}(\Gamma))$, $0 < \alpha < 1$. Now, from the continuous embedding of $H^{2\alpha}$ into $H^{2\alpha-\varepsilon}$, and by applying the Arzelà-Ascoli theorem, we conclude that $(\bar{\eta}_N)_{N\in\mathbb{N}}$ has a convergent subsequence, which we again denote by $(\bar{\eta}_N)_{N\in\mathbb{N}}$, such that

$$\bar{\eta}_N \to \eta \text{ in } C([0, T]; H^s(\Gamma)), \quad 0 < s < 2.$$

Here, we used the fact that the sequences $(\bar{\eta}_N)_{N\in\mathbb{N}}$ i $(\eta_N)_{N\in\mathbb{N}}$ have the same limit. We now prove the following result:

Lemma 2.3 $\eta_N \to \eta$ *in* $L^\infty(0, T; H^s(\Gamma))$, $0 < s < 2$.

Proof The proof follows from the continuity in time of η, and from the fact that $\bar{\eta}_N \to \eta$ in $C([0, T]; H^s(\Gamma))$, $0 < s < 2$. Namely, let $\varepsilon > 0$. Then, from the continuity of η in time we have that there exists a $\delta t > 0$ such that

$$\|\eta(t_1) - \eta(t_2)\|_{H^s(\Gamma)} < \frac{\varepsilon}{2}, \text{ for } t_1, t_2 \in [0, T], \text{ and } |t_1 - t_2| \leq \delta t.$$

Furthermore, from the convergence $\bar{\eta}_N \to \eta$ in $C([0, T]; H^s(\Gamma))$, $0 < s < 2$, we know that there exists an $N^* \in \mathbb{N}$ such that

$$\|\bar{\eta}_N - \eta\|_{C([0,T]; H^s(\Gamma))} < \frac{\varepsilon}{2}, \ \forall N \geq N^*.$$

Now, let N be any natural number such that $N > \max\{N^*, T/\delta t\}$. Denote by $\Delta t = T/N$, and let $t \in [0, T]$. Furthermore, let $n \in \mathbb{N}$ be such that $(n - 1)\Delta t < t \leq n\Delta t$. Recall that $\bar{\eta}_N(n\Delta t) = \eta_N(n\Delta t) = \eta_N(t)$ from the definition of $\bar{\eta}_N$ and η_N. By using this, and by combining the two estimates above, we get that for $(n - 1)\Delta t < t \leq n\Delta t$:

$$\|\eta_N(t) - \eta(t)\|_{H^s(\Gamma)} = \|\eta_N(t) - \eta(n\Delta t) + \eta(n\Delta t) - \eta(t)\|_{H^s(\Gamma)}$$

$$= \|\eta_N(n\Delta t) - \eta(n\Delta t) + \eta(n\Delta t) - \eta(t)\|_{H^s(\Gamma)}$$

$$\leq \|\eta_N(n\Delta t) - \eta(n\Delta t)\| + \|\eta(n\Delta t) - \eta(t)\|_{H^s(\Gamma)}$$

$$= \|\bar{\eta}_N(n\Delta t) - \eta(n\Delta t)\|_{H^s(\Gamma)} + \|\eta(n\Delta t) - \eta(t)\|_{H^s(\Gamma)} < \varepsilon.$$

Here, the first term is bounded by $\varepsilon/2$ due to the convergence $\bar{\eta}_N \to \eta$, while the second term is bounded by $\varepsilon/2$ due to the continuity of η. Since the obtained estimate is uniform in N and t, the statement of the Lemma is proved.

We conclude this section by showing one last convergence result that will be used in the next section to prove that the limiting functions satisfy the weak formulation of the FSI problem.

Lemma 2.4 *Structure displacements* $(\eta_N)_{N \in \mathbb{N}}$ *are such that:*

$$\begin{aligned} \eta_N &\to \eta \text{ in } L^\infty(0, T; C^1[0, L]), \\ \tau_{\Delta t}\eta_N &\to \eta \text{ in } L^\infty(0, T; C^1[0, L]). \end{aligned} \tag{2.6.2}$$

Proof The first statement is a direct consequence of Lemma 2.3 in which we proved that $\eta_N \to \eta$ in $L^\infty(0, T; H^s(0, L))$, $0 < s < 2$. This means that for $s > \frac{3}{2}$ we immediately have

$$\eta_N \to \eta \text{ in } L^\infty(0, T; C^1[0, L]). \tag{2.6.3}$$

To show convergence of the shifted displacements $\tau_{\Delta t}\eta_N$ to the same limiting function η, we recall that

$$\bar{\eta}_N \to \eta \text{ in } C([0, T]; H^s[0, L]), \ 0 < s < 2,$$

and that $(\bar{\eta}_N)_{N \in \mathbb{N}}$ is uniformly bounded in $C^{0,1-\alpha}([0, T]; H^{2\alpha}(0, L)), 0 < \alpha < 1$. Uniform boundeness of $(\bar{\eta}_N)_{N \in \mathbb{N}}$ in $C^{0,1-\alpha}([0, T]; H^{2\alpha}(0, L))$ implies that there exists a constant $C > 0$, independent of N, such that

$$\|\bar{\eta}_N((n-1)\Delta t) - \bar{\eta}_N(n\Delta t)\|_{H^{2\alpha}(0,L)} \leq C|\Delta t|^{1-\alpha}.$$

This means that for each $\varepsilon > 0$, there exists an $N_1 > 0$ such that

$$\|\bar{\eta}_N((n-1)\Delta t) - \bar{\eta}_N(n\Delta t)\|_{H^{2\alpha}(0,L)} \leq \frac{\varepsilon}{2}, \text{ for all } N \geq N_1.$$

Here, N_1 is chosen by recalling that $\Delta t = T/N$, and so the right hand-side implies that we want an N_1 such that

$$C\left(\frac{T}{N}\right)^{1-\alpha} < \frac{\varepsilon}{2} \text{ for all } N \geq N_1.$$

Now, convergence $\bar{\eta}_N \to \eta$ in $C([0, T]; H^s[0, L]), \ 0 < s < 2$, implies that for each $\varepsilon > 0$, there exists an $N_2 > 0$ such that

$$\|\bar{\eta}_N(n\Delta t) - \eta(t)\|_{H^s(0,L)} < \frac{\varepsilon}{2}, \text{ for all } N \geq N_2.$$

We will use this to show that for each $\varepsilon > 0$ there exists an $N^* \geq \max\{N_1, N_2\}$, such that

$$\|\tau_{\Delta t}\bar{\eta}_N(t) - \eta(t)\|_{H^s(0,L)} < \varepsilon, \text{ for all } N \geq N^*.$$

Let $t \in (0, T)$. Then there exists an n such that $t \in ((n-1)\Delta t, n\Delta t]$. We calculate

$$\|\tau_{\Delta t}\bar{\eta}_N(t) - \eta(t)\|_{H^s(0,L)} = \|\tau_{\Delta t}\bar{\eta}_N(t) - \bar{\eta}_N(n\Delta t) + \bar{\eta}_N(n\Delta t) - \eta(t)\|_{H^s(0,L)}$$

$$= \|\bar{\eta}_N((n-1)\Delta t) - \bar{\eta}_N(n\Delta t) + \bar{\eta}_N(n\Delta t) - \eta(t)\|_{H^s(0,L)}$$

$$\leq \|\bar{\eta}_N((n-1)\Delta t) - \bar{\eta}_N(n\Delta t)\|_{H^s(0,L)} + \|\bar{\eta}_N(n\Delta t) - \eta(t)\|_{H^s(0,L)}.$$

The first term is less than ε for all $N > N^*$ by the uniform boundeness of $(\bar{\eta}_N)_{N \in \mathbb{N}}$ in $C^{0,1-\alpha}([0, T]; H^{2\alpha}(0, L))$, while the second term is less than ε for all $N > N^*$ by the convergence of $\bar{\eta}_N$ to η in $C([0, T]; H^s[0, L]), \ 0 < s < 2$.

Now, by noticing that $\tau_{\Delta t}\bar{\eta}_N = \overline{(\tau_{\Delta t}\eta_N)}$ we can use the same argument as in Lemma 2.2 to show that sequences $\overline{(\tau_{\Delta t}\eta_N)}$ and $\tau_{\Delta t}\eta_N$ both converge to the same limit η in $L^\infty(0, T; H^s(0, L))$, for $s < 2$.

2.6.2 Strong Convergence of Fluid and Structure Velocities

We would like to show that the sequences $(\mathbf{u}_N)_{N\in\mathbb{N}}$ and $(v_N)_{N\in\mathbb{N}}$, constructed above, are relatively compact in $L^2(0, T; L^2(\Omega))$ and $L^2(0, T; L^2(\Gamma))$, respectively. This will be sufficient to pass to the limit in (2.4.15) and show that the limits satisfy the weak formulation (2.3.12).

Remark 2.7 (Back to the Physical Space) To show relative compactness of the fluid and structure velocities, it is more convenient to work in the physical space, namely on $\Omega^\eta(t)$, instead of the fixed domain Ω.

On $\Omega^\eta(t)$, the weak formulation of the coupled semi-discretized problem is obtained by transforming the weak formulation (2.4.15) onto the moving domain. Similarly, the function spaces introduced in (2.4.12) and used in the definition of the semi-discretized, coupled problem defined on the fixed domain Ω, are transformed into the following function spaces defined on domains $\Omega^{\eta_N^n}$:

$$(\mathcal{V}_F^\eta)_N^{n+1}(\Omega^{\eta_N^n}) = \{(\mathbf{u}^{n+1}, v^{n+1}) \in \mathcal{V}_F(\Omega^{\eta_N^n}) \times L^2(\Gamma) \ : \ \mathbf{u}^{n+1}_{|\Gamma^{\eta_N^n}} = v^{n+1}\mathbf{e}_r\},$$
$$(\mathcal{Q}^\eta)_N^n(\Omega^{\eta_N^n}) = \{(\mathbf{q}, \psi) \in \mathcal{V}_F(\Omega^{\eta_N^n}) \times H_0^2(\Gamma) \ : \ \mathbf{q}_{|\Gamma^{\eta_N^n}} = \psi\mathbf{e}_r\}.$$
$$(2.6.4)$$

The weak formulation of the *semi-discretized, coupled* FSI problem on moving domains $\Omega^{\eta_N^n}$ reads: find $(\mathbf{u}_N^{n+1}, v_N^{n+1}) \in (\mathcal{V}_F^\eta)_N^{n+1}(\Omega^{\eta_N^n})$ and $(v_N^{n+\frac{1}{2}}, \eta_N^{n+\frac{1}{2}}) \in H_0^2(\Gamma) \times H_0^2(\Gamma)$ such that:

$$\varrho_f \int_{\Omega^{\eta_N^n}} \frac{\mathbf{u}_N^{n+1} - \widehat{\mathbf{u}}_N^n}{\Delta t} \cdot \mathbf{q} + \frac{\rho_f}{2} \int_{\Omega^{\eta_N^n}} \frac{v_N^{n+1/2}}{R+\eta_N^n}\mathbf{u}_N^{n+1} \cdot \mathbf{q} + 2\mu \int_{\Omega^{\eta_N^n}} \mathbf{D}(\mathbf{u}_N^{n+1}) : \mathbf{D}(\mathbf{q})$$

$$+\varrho_f \int_{\Omega^{\eta_N^n}} \frac{1}{2}\left[(\widehat{\mathbf{u}}_N^n - \frac{v_N^{n+1/2}r}{R+\eta_N^n}\mathbf{e}_r) \cdot \nabla \mathbf{u}_N^{n+1} \cdot \mathbf{q} - (\widehat{\mathbf{u}}_N^n - \frac{v_N^{n+1/2}}{R+\eta_N^n}\mathbf{e}_r) \cdot \nabla \right] \mathbf{q} \cdot \mathbf{u}_N^{n+1}$$

$$+\rho_s h \int_\Gamma \frac{v_N^{n+1} - v_N^n}{\Delta t}\psi + a_e(\eta_N^{n+1}, \psi) = F^n(\mathbf{q}),$$

$$\int_\Gamma \frac{\eta_N^{n+1} - \eta_N^n}{\Delta t}\psi = \int_\Gamma v_N^{n+1/2}\psi, \quad \forall(\mathbf{q}, \psi) \in \mathcal{Q}_N^n(\Omega^{\eta_N^n}),$$
$$(2.6.5)$$

$$\text{with } \nabla \cdot \mathbf{u}_N^{n+1} = 0.$$

Here:

$$\widehat{\mathbf{u}}_N^n = \mathbf{u}_N^n \circ \mathcal{A}^{\eta_N^{n-1}} \circ (\mathcal{A}^{\eta_N^n})^{-1}.$$
$$(2.6.6)$$

Thus, $\hat{\mathbf{u}}_N^n$, which is defined on $\Omega^{\eta_N^n}$, is obtained by taking the function $\mathbf{u}_N^n$, which is defined on $\Omega^{\eta_N^{n-1}}$, mapping it back onto Ω via $\mathcal{A}^{\eta_N^{n-1}}$, and then composing it with $(\mathcal{A}^{\eta_N^n})^{-1}$ to get the function defined on $\Omega^{\eta_N^n}$.

Stronger Velocity Solution Space To obtain the desired compactness result, we consider the integral formulation (2.6.5), but defined on a slightly different, "stronger" function spaces. For each fixed N we introduce

$$(V_F^\eta)_N^{n+1}(\Omega^{\eta_N^n}) = \{(\mathbf{u}, v) \in V_F(\Omega^{\eta_N^n}) \times H^{1/2}(\Gamma) : \mathbf{u}_{|\Gamma^{\eta_N^n}} = v\} \tag{2.6.7}$$

which is a solution space which utilizes the fact that our weak solutions have the trace of the fluid velocity not only in $L^2(\Gamma)$ but also in $H^{1/2}(\Gamma)$. Similarly, we introduce the test space

$$(Q^\eta)_N^n(\Omega^{\eta_N^n}) = \{(\mathbf{q}, \psi) \in \left(V_F(\Omega^{\eta_N^n}) \cap H^3(\Omega^{\eta_N^n})\right) \times H_0^2(\Gamma) : \mathbf{q}_{|\Gamma^{\eta_N^n}} = \psi\}, \tag{2.6.8}$$

by requiring that the fluid velocity test functions additionally belong to $H^3(\Omega^{\eta_N^n})$. The resulting test space in dense in $(Q^\eta)_N^n(\Omega^{\eta_N^n})$. This additional regularity simplifies some calculations in the compactness proof (although it is not optimal in the sense that weaker assumptions could be used in exchange for a slightly more complicated estimates).

To simplify notation, without loss of generality, in the rest of this section we will be taking all the physical constants to be equal to 1.

2.6.2.1 Extensions to the Maximal Domain

One of the crucial issues in designing a compactness argument for this class of problems is how to deal with the time-dependent motion of fluid domains (see Fig. 2.5). More precisely, the question is how to compare sequences of approximate

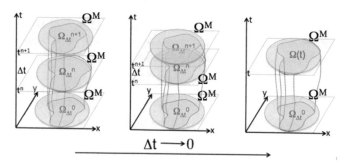

Fig. 2.5 A sketch of moving domains $\Omega_N^{\eta_N^n}$ and the "maximal domain" Ω_M

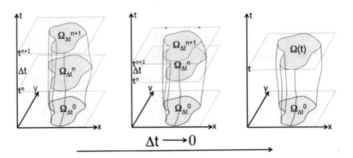

Fig. 2.6 A sketch of moving domains $\Omega_N^{\eta^n}$ as they change with $\Delta t \to 0$, i.e., with $N \to \infty$

solutions which are defined on different fluid domains, see Fig. 2.6. Even more generally, the question is how to design a compactness argument for a family of problems:

$$\frac{dU}{dt} = A^t U, \quad t \in (0, T),$$
$$U(0) = U_0.$$

where operators A^t are defined on functions spaces that depend on time (via fluid domain motion).

To deal with this difficulty, we will introduce a domain Ω^M which contains all the fluid domains Ω^{η^n}, and extend the fluid velocities and the corresponding test functions to Ω^M. The following Lemma will guarantee the existence of such a domain. Moreover, it will provide the necessary estimates that will be useful in proving compactness in time.

Lemma 2.5 *There exist smooth functions $m(z)$ and $M(z)$, $z \in \overline{\Gamma} = [0, L]$, such that the sequence $\{\eta_N^n(z)\}$ corresponding to the semi-discretization of $(0, T)$ can be uniformly bounded as follows:*

- *$m(z) \le \eta_N^n(z) \le M(z)$, $z \in [0, L]$, $n = 0, \dots, N$, $N \in \mathbb{N}$, $\Delta t = T/N$, and*
- *$M(z) - m(z) \le C(T)$, $z \in [0, L]$, where $C(T) \to 0$, $T \to 0$.*

Moreover, for each fixed N, $l \in \{1, \dots, N\}$, and $n \in \{1, \dots, N - 1\}$, there exist smooth functions $m_N^{n,l}(z)$, $M_N^{n,l}(z)$ such that the functions $\{\eta_N^n(z)\}$ corresponding to the semi-discretization of $(t, t+h) = (n\Delta t, (n+l)\Delta t)$ can be bounded as follows:

- *$m_N^{n,l}(z) \le \eta_N^{n+i}(z) \le M_N^{n,l}(z)$, $z \in [0, L]$, $i = 0, \dots, l$, and*
- *$M_N^{n,l}(z) - m_N^{n,l}(z) \le C\sqrt{l\Delta t}$, $z \in [0, L]$, and $\|M_N^{n,l} - m_N^{n,l}\|_{L^2(\Gamma)} \le Cl\Delta t$.*

Proof The existence and properties of the functions M and m are a direct consequence of Proposition 5 in [62, p. 29].

To prove the second statement of Lemma 2.5, we fix $N \in \mathbb{N}$ and consider finitely many functions $\eta_N^{n+i}(z)$ for $i = 0, \dots, l$, which are defined on the time interval from $t = n\Delta t$ to $t + h = n\Delta t + l\Delta t$. Define $m_N^{n,l}(z)$ and $M_N^{n,l}(z)$, $z \in [0, L]$ to be the functions obtained by considering the minimum and maximum of the finitely many functions $\eta_N^{n+i}(z)$ for $i = 0, \ldots, l$, mollified if necessary to get the smooth functions. The properties of the functions $m_N^{n,l}$, $M_N^{n,l}$ then follow from the proof of the same Proposition 5 in [62]. Namely, from the proof of Proposition 5 [62] it follows that the upper bound on $\|\eta_N^{n+i} - \eta_N^n\|_{H^1}$ only depends on the width of the time interval, which is h, namely $\|\eta_N^{n+i} - \eta_N^n\|_{H^1} \leq C\sqrt{h}$, for all $i = 1, \ldots, l$. Since $\overline{\eta_N^n}$ are defined on $\overline{\Gamma} = [0, L]$, then the estimate also holds point-wise (because of the Sobolev embedding of $H^1(\Gamma)$ into $C[0, L]$). Thus the first inequality is proven since $h = l\Delta t$. The second inequality, namely the L^2-bound on the difference between $M_N^{n,l}$ and $m_N^{n,l}$, follows from

$$\|\eta_N^n - \eta_0\|_{L^2(\Gamma)} \leq \sum_{i=0}^{n-1} \|\eta_N^{i+1} - \eta_N^i\|_{L^2(\Gamma)} = \Delta t \sum_{i=0}^{n-1} \|v_N^{i+\frac{1}{2}}\|_{L^2(\Gamma)} \leq \|v_N\|_{L^2(0,T,L^2(\Gamma))},$$

where $\eta_N^0 = \eta_0$, and from the uniform energy bound of the structure velocity v_N in $L^2(0, T; L^2(\Gamma))$, which follows from Lemma 2.1.

We consider the functions $\mathbf{u}_N^n$ defined on $\Omega^{\eta_N^{n-1}}$, and the approximations v_N^n defined on Γ. We define **extensions** of the functions $\mathbf{u}_N^n$ to the maximal domain Ω^M to be the functions $\mathbf{u}_N^n$ extended by zero to Ω^M. With a slight abuse of notation, we use the same notation to denote the extended functions.

Lemma 2.6 *The extensions by 0 to Ω^M of $\mathbf{u}_N^n$ are such that*

$$(\mathbf{u}_N^n, v_N^n) \in H^s(\Omega^M) \times H^s(\Gamma), 0 < s < 1/2.$$

Proof The proof is a consequence of Theorem 2.7 in [60], which states that for Lipschitz domains, extensions by 0 of H^1-functions belong to H^s, $0 < s < 2$. More precisely, to apply Theorem 2.7 in [60] we first notice that the functions η_N^n are uniformly Lipschitz on Γ. This follows from the uniform energy estimates presented in Lemma 2.1 above, which imply that there exists a constant $C > 0$, independent of Δt, such that the $H^2(\Gamma)$-norms of η_N^n are uniformly bounded by C (second estimate in statement e of Lemma 2.1). Since $\Gamma \subset \mathbb{R}$, this implies that η_N^n are uniformly Lipschitz, where the Lipschitz constant is independent of Δt. Now we can apply Theorem 2.7 in [60] to the functions $\mathbf{u}_N^n$. First, the uniform energy estimates from Lemma 2.1 imply that $\mathbf{u}_N^n$ are in $H^1(\Omega^{\eta_N^n})$. Then, Theorem 2.7 in [60] implies that the extensions by 0 to Ω^M belong to $H^s(\Omega^M)$, $0 < s < 1/2$.

Moreover, there exists an absolute constant $C > 0$, which depends only on the maximum of the Lipschitz constants for η_N^n, such that

$$\|\mathbf{u}_N^n\|_{H^s(\Omega^M)} \leq C\|\nabla \mathbf{u}_N^n\|_{L^2(\Omega^{\eta_N^n})}, 0 < s < 1/2. \tag{2.6.9}$$

In 1D the constant C "measures" the jump across Γ.

Similarly, we extend the test functions $\mathbf{q} \in Q_N^n$ to the maximal domain Ω^M by 0.

Remark 2.8 (Extensions of Test Functions) Notice that the extended test functions are no longer smooth. However, as we shall see below, we will work with "admissible" test functions whose smoothness will depend only on the smoothness of $\mathbf{q}$ within the domains $\Omega^{\eta_N^n}$.

We are now in a position to apply the following compactness result to our approximate sequences of velocity functions.

2.6.2.2 A Generalization of the Aubin–Lions–Simon Compactness Lemma to Problems on Moving Domains [67]

We state a generalization of Aubin–Lions–Simon compactness lemma, recently obtained by Muha and Čanić in [67], which will be used to prove precompactness of the fluid and structure velocities in $L^2(0, T; H)$, where H is a Hilbert space specified below in Sect. 2.6.2.3.

To show compactness in time, it is useful to introduce the definition of time-shifts: for an arbitrary function $\mathbf{f}$ denote by $\tau_h \mathbf{f}$ the time-shifts

$$\tau_h \mathbf{f}(t, .) = \mathbf{f}(t - h, .), \quad h \in \mathbb{R}. \tag{2.6.10}$$

Theorem 2.1 ([67]) *Let V and H be Hilbert spaces such that $V \subset\subset H$. Suppose that $\{\mathbf{u}_N\} \subset L^2(0, T; H)$ is a sequence such that $\mathbf{u}_N(t, \cdot) = \mathbf{u}_N^n(\cdot)$ on $((n-1)\Delta t, n\Delta t]$, $n = 1, \ldots, N$, with $N\Delta t = T$. Let V_N^n and Q_N^n be Hilbert spaces such that $(V_N^n, Q_N^n) \hookrightarrow V \times V$, where the embeddings are uniformly continuous w.r.t. N and n, and $V_N^n \subset\subset \overline{Q_N^n}^H \hookrightarrow (Q_N^n)'$. Let $\mathbf{u}_N^n \in V_N^n, n = 1, \ldots, N$. Suppose that the following is true:*

(A) There exists a universal constant $C > 0$ such that for every $\Delta t > 0$

(A1) $\|\mathbf{u}_N\|_{L^2(0,T;V)} \leq C,$

(A2) $\|\mathbf{u}_N\|_{L^\infty(0,T;H)} \leq C,$

(A3) $\|\tau_{\Delta t}\mathbf{u}_N - \mathbf{u}_N\|_{L^2(\Delta t, T; H)}^2 \leq C\Delta t.$

(B) There exists a universal constant $C > 0$ such that

$$\left\| P_N^n \frac{\mathbf{u}_N^{n+1} - \mathbf{u}_N^n}{\Delta t} \right\|_{(Q_N^n)'} \leq C(\|\mathbf{u}_N^{n+1}\|_{V_N^{n+1}} + 1), \quad n = 0, \ldots, N - 1,$$

where P_N^n is the orthogonal projector onto $\overline{Q_N^n}^H$.

(C) The function spaces Q_N^n and V_N^n depend smoothly on time in the following sense:

(C1) For every $\Delta t > 0$, and for every $l \in \{1, \ldots, N\}$ and $n \in \{1, \ldots, N - 1\}$, there exists a space $Q_N^{n,l} \subset V$ and the operators $J_{N,l,n}^i : Q_N^{n,l} \to$

Q_N^{n+i}, $i = 0, 1, \ldots, l$, such that $\|J_{N,l,n}^i \mathbf{q}\|_{Q_N^{n+i}} \le C\|\mathbf{q}\|_{Q_N^{n,l}}$, $\forall \mathbf{q} \in Q_N^{n,l}$, and

$$\|J_{N,l,n}^i \mathbf{q} - J_{N,l,n}^j \mathbf{q}\|_H \le C|i-j|\Delta t \|\mathbf{q}\|_{Q_N^{n,l}}, \quad i, j \in \{0, \ldots, l\},$$

(2.6.11)

$$\|J_{N,l,n}^i \mathbf{q} - \mathbf{q}\|_H \le C\sqrt{l\Delta t}\|\mathbf{q}\|_{Q_N^{n,l}}, \quad i \in \{0, \ldots, l\},$$

(2.6.12)

where $C > 0$ is independent of $\Delta t(N)$, n and l.

(C2) Let $V_N^{n,l} = \overline{Q_N^{n,l}}^V$. There exist the functions $I_{N,l,n}^i : V_N^{n+i} \to V_N^{n,l}$, $i = 0, 1, \ldots, l$, and a universal constant $C > 0$, such that for every $\mathbf{v} \in V_N^{n+i}$

$$\|I_{N,l,n}^i \mathbf{v}\|_{V_N^{n,l}} \le C\|\mathbf{v}\|_{V_N^{n+i}}, \quad i \in \{0, \ldots, l\},$$

(2.6.13)

$$\|I_{N,l,n}^i \mathbf{v} - \mathbf{v}\|_H \le g(l\Delta t)\|\mathbf{v}\|_{V_N^{n+i}}, \quad i \in \{0, \ldots, l\},$$

(2.6.14)

where $g : \mathbb{R}_+ \to \mathbb{R}_+$ is a universal, monotonically increasing function such that $g(h) \to 0$ as $h \to 0$.

(C3) Uniform Ehrling property: For every $\delta > 0$ there exists a constant $C(\delta)$ independent of n, l and $\Delta t(N)$, such that

$$\|\mathbf{v}\|_H \le \delta\|\mathbf{v}\|_{V_N^{n,l}} + C(\delta)\|\mathbf{v}\|_{(Q_N^{n,l})'}, \quad \mathbf{v} \in V_N^{n,l}.$$

(2.6.15)

Then $\{\mathbf{u}_N\}_{N \in \mathbb{N}}$ is relatively compact in $L^2(0, T; H)$.

Remark 2.9 Conditions (A) and (B) correspond the "classical" conditions in the Aubin-Lions lemma for problems on *fixed domains* [34], with $V_N^n = V$, $Q_N^{n,l} = Q_N^n = Q$. Namely, in that case condition (C) is trivially satisfied. Condition (A3) is not necessary, however, it is usually satisfied as a by-product of the time-discretization approach (Rothe's method), see [67]. Conditions (C) are crucial for compactness arguments on moving domains. They are directly related to making sense of taking the time derivative on different domains. Conditions (C1) and (C2) are "local" in the sense that they require smooth dependence of the test spaces and solution spaces on time, locally in time, namely, for the time shifts from $n\Delta t$ to time $(n + l)\Delta t$. Condition (C3), i.e., the Uniform Ehrling Property, provides a "global" estimate. More precisely, while the classical Ehrling lemma holds for any set of function spaces parameterized by l, n and Δt, i.e., $\{V_N^{n,l}, H_N^{n,l}, (Q_N^{n,l})'\} = \{V_N^{n,l}, \overline{Q_N^{n,l}}^H, (Q_N^{n,l})'\}$, $V_N^{n,l} \hookrightarrow H_N^{n,l} \simeq (H_N^{n,l})' \hookrightarrow (Q_N^{n,l})'$ where the first embedding is compact and the last is injective, with the constants $C(\delta)$ depending on l, n and Δt, the Uniform Ehrling Property, i.e., (C3) requires that those constants *do not depend* on the parameters n, l and N, which is a condition on the entire family of function spaces. Thus, the Uniform Ehrling Property requires that the classical Ehrling lemma holds *uniformly* for the entire family of function spaces $\{V_N^{n,l}, H_N^{n,l}, (Q_N^{n,l})'\}$ parameterized by l, n and N.

Condition (C1) states that there exists a "common" test space $Q_N^{n,l}$, which is continuously embedded in all the test spaces $Q_N^{n,i}, i = 1, \ldots, l$ for the time shifts $(n+i)\Delta t, i = 1, \ldots, l$, and such that properties (2.6.11) and (2.6.12) hold on $Q_N^{n,l}$, which is related to making sense of approximate time derivatives.

Condition (C2) is similar in the sense that it states certain properties of a "common" solution space $V_N^{n,l}$, common for the time shifts $(n+i)\Delta t, i = 1, \ldots, l$, satisfying properties (2.6.13) and (2.6.14), which state continuous dependence of the solutions spaces on time, locally near time $n\Delta t$, for an arbitrary n.

2.6.2.3 Compactness of Approximations for Fluid and Structure Velocities

We will now apply the compactness result in Theorem 2.1 to the sequences of approximate fluid and structure velocities. We first define the overarching Hilbert spaces H and V, which do not depend on N, and satisfy the properties listed in Theorem 2.1. For this purpose we recall the definition of the maximal fluid domain Ω_M, and consider the fluid velocity functions $\mathbf{u}_N^n$ extended to Ω_M by zero. By Lemma 2.6, the functions $(\mathbf{u}_N^n, v_N^n)$ belong to the space $H^s(\Omega^M) \times H^s(\Gamma), 0 < s < 1/2$. Keeping this result in mind, we define the spaces H and V of functions $(\mathbf{u}_N^n, v_N^n)$ as follows:

$$H = L^2(\Omega^M) \times L^2(\Gamma), \quad V = H^s(\Omega^M) \times H^s(\Gamma), 0 < s < 1/2. \quad (2.6.16)$$

Obviously, H and V are Hilbert spaces and $V \subset\subset H$.

Furthermore, let V_N^n and Q_N^n from Theorem 2.1 be given by the spaces (2.6.7) and (2.6.8), respectively:

$$V_N^n := (V_F^\eta)_N^{n+1}(\Omega^{\eta_N^n}), \quad Q_N^n := (Q^\eta)_N^{n+1}(\Omega^{\eta_N^n}).$$

Then, from inequality (2.6.9) we see that V_N^n is continuously embedded in V, uniformly in N and n. Furthermore, $V_N^n \subset\subset \overline{Q_N^n}^H \hookrightarrow (Q_N^n)'$.

Now, define the velocity functions which depend on both time and space by introducing $(\mathbf{u}_N, v_N) \subset L^2(0, T; H)$ which are piecewise constant in t, i.e.

$$\left.\begin{array}{l} \mathbf{u}_N = \mathbf{u}_N^n \\ v_N = v_N^n \end{array}\right\} \quad \text{on } ((n-1)\Delta t, n\Delta t], \ n = 1, \ldots, N,$$

as well as the corresponding time-shifts, denoted by τ_h, defined by

$$\tau_h \mathbf{u}_N(t, .) = \mathbf{u}_N(t - h, .), \ h \in \mathbb{R}.$$

Theorem 2.2 (Compactness of $(\mathbf{u}_N, v_N)_{N \in \mathbb{N}}$) *The sequence* $(\mathbf{u}_N, v_N)_{N \in \mathbb{N}}$, *satisfying the weak formulation* (2.6.5) *and energy estimates from Lemma 2.1, is relatively compact in* $L^2(0, T; H)$, *where* $H = L^2(\Omega^M)^2 \times L^2(\Gamma)$.

Proof We would like to show that the assumptions (A)-s-(C) from Theorem 2.1 hold true.

Property (A): Strong Bounds The strong bounds (A1) and (A2) follow directly from the uniform energy bounds in Lemma 2.1 and from Lemma 2.6 above. As stated before, property (A3) is not needed for the proof.

Property (B): The Time Derivative Bound We want to estimate the following norm:

$$\left\| P_N^n \frac{\mathbf{u}_N^{n+1} - \mathbf{u}_N^n}{\Delta t} \right\|_{(Q_N^n)'} = \sup_{\|(\mathbf{q}, \psi)\|_{Q_N^n} = 1} \left| \int_{\Omega^{\eta_N^n}} \frac{\mathbf{u}_N^{n+1} - \mathbf{u}_N^n}{\Delta t} \cdot \mathbf{q} + \int_0^L \frac{v_N^{n+1} - v_N^n}{\Delta t} \psi \right|.$$

For this purpose we add and subtract the function $\widetilde{\mathbf{u}}_N^n$ which is defined in (2.6.6):

$$\left| \int_{\Omega^{\eta_N^n}} \frac{\mathbf{u}_N^{n+1} - \mathbf{u}_N^n \pm \widetilde{\mathbf{u}}_N^n}{\Delta t} \cdot \mathbf{q} + \int_0^L \frac{v_N^{n+1} - v_N^n}{\Delta t} \psi \right| \qquad (2.6.17)$$

$$\leq \left| \int_{\Omega^{\eta_N^n}} \frac{\mathbf{u}_N^{n+1} - \widetilde{\mathbf{u}}_N^n}{\Delta t} \cdot \mathbf{q} + \int_0^L \frac{v_N^{n+1} - v_N^n}{\Delta t} \psi \right| + \left| \int_{\Omega^{\eta_N^n}} \frac{\widetilde{\mathbf{u}}_N^n - \mathbf{u}_N^n}{\Delta t} \cdot \mathbf{q} \right|,$$

and estimate the two terms on the right hand-side.

The first term is estimated by using the weak form of the problem given by Eq. (2.6.5):

$$\left| \int_{\Omega^{\eta_N^n}} \frac{\mathbf{u}_N^{n+1} - \widehat{\mathbf{u}}_N^n}{\Delta t} \cdot \mathbf{q} + \int_0^L \frac{v_N^{n+1} - v_N^n}{\Delta t} \psi \right|$$

$$\leq C \frac{R + M}{R + m} \|\nabla \mathbf{q}\|_{L^\infty} (\|v_N^{n+1/2}\|_{L^2} + \|\mathbf{u}_N^n\|_{L^2}) \|\nabla \mathbf{u}_N^{n+1}\|_{L^2} + C_1 \|\nabla \mathbf{u}_N^{n+1}\|_{L^2} \|\nabla \mathbf{q}\|_{L^2}$$

$$+ \|\eta_N^{n+1}\|_{H^2} \|\psi\|_{H^2} + C\|\mathbf{q}\|_{H^1} \leq C(\|\nabla \mathbf{u}_N^{n+1}\|_{L^2} + \|\eta_N^{n+1}\|_{H^2} + 1) \|(\mathbf{q}, \psi)\|_{(Q_N^{n,l})}.$$

Here we used the energy estimates from Lemma 2.1 from where we concluded that $\|\mathbf{u}_N^n\|_{L^2}$, $\|v_N^n\|_{L^2}$ are uniformly bounded by C.

Remark 2.10 Notice how the choice of the space Q_N^n, which includes high regularity Sobolev spaces, is useful in the last inequality to provide the upper bound in terms of $\|(\mathbf{q}, \psi)\|_{(Q_N^n)}$.

To estimate the second term, we first notice that function $\widehat{\mathbf{u}}_N^n$ is non-zero on $\Omega^{\eta_N^n}$, while function $\mathbf{u}_N^n$ is non-zero on $\Omega^{\eta_N^{n-1}}$. This is why we introduce $A = \Omega^{\eta_N^{n-1}} \cap \Omega^{\eta^n}$,

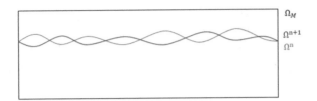

Fig. 2.7 The 2D fluid domains at time steps t_n and t_{n+1}

$B_1 = \Omega^{\eta^n} \setminus \Omega^{\eta^{n-1}_N}$, and $B_2 = \Omega^{\eta^{n-1}_N} \setminus \Omega^{\eta^n}$, see Fig. 2.7, and estimate the integrals over A, B_1, and B_2 separately:

$$
|\int_A (\widetilde{\mathbf{u}}^n_N - \mathbf{u}^n_N) \cdot \mathbf{q}| = |\int_A \left(\mathbf{u}^n_N(z, r) - \mathbf{u}^n_N(z, \frac{R + \eta^n_N}{R + \eta^{n-1}_N} r)\right) \cdot \mathbf{q}(z, r) dz dr|
$$

$$
\leq C \Delta t \|v^{n-1/2}\|_{L^2} \|\nabla \mathbf{u}^n_N\|_{L^2(A)} \|\mathbf{q}\|_{L^\infty(A)}.
$$

To estimate the integral over B_1, we use the fact that $\mathbf{u}^n_N = 0$ on B_1 to obtain:

$$
|\int_{B_1} (\widetilde{\mathbf{u}}^n_N - \mathbf{u}^n_N) \cdot \mathbf{q}| \leq |\int_0^L dz \int_{R+\eta^{n-1}_N(z)}^{R+\eta^n_N(z)} (\widetilde{\mathbf{u}}^n_N - \mathbf{u}^n_N)(z, r) \cdot \mathbf{q} dr|
$$

$$
= \int_0^L dz \int_{R+\eta^{n-1}_N(z)}^{R+\eta^n_N(z)} \widetilde{\mathbf{u}}^n_N(z, r) \cdot \mathbf{q} dr|
$$

$$
\leq \|\mathbf{q}\|_{L^\infty} \int_0^L \max_r (\widetilde{\mathbf{u}}^n_N(z, r)) dz \int_{R+\eta^{n-i}_N(z)}^{R+\eta^n_N(z)} dr
$$

$$
\leq C \|\mathbf{q}\|_{L^\infty} \int_0^L \|\partial_r u^n_{N,r}(., z)\|_{L^2_r} |\Delta t v^{n-1/2}(z)| dz
$$

$$
\leq C \Delta t \|\mathbf{q}\|_{L^\infty} \|\nabla \mathbf{u}^n_N\|_{L^2(\Omega^n_N)}.
$$

Here, we used the fact that for $f \in H^1(0, 1)$ we can estimate $\|f\|_{L^\infty} \leq C \|f'\|_{L^2}$, and applied this to the function $\mathbf{u}^n_N$ above, viewed as a function of only one variable, r, satisfying the condition that the r-component $u_r(z, 0) = 0$ is equal to zero at $r = 0$, and the z-component u_z is equal to zero on the boundary $r = R + \eta^{n-1}_N(z)$. This is a formal estimate which can be justified by density arguments.

The integral over B_2 is equal to zero.

Property (C): Smooth Dependence of Function Spaces on Time We want to define common function spaces for test functions and for approximate solutions, which will help with estimating the time shifts. For this purpose we recall that our scheme is designed in such a way that the functions $\mathbf{u}^{n+1}_N$ are defined on the "previous" domain $\Omega^{\eta^n_N}$.

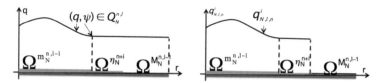

Fig. 2.8 A sketch of the test functions $\mathbf{q} \in Q_N^{n,l}$ (left) and $\mathbf{q}_{N,l,n}^i$ (right) as functions of r only

Property C1: The Common Test Space Consider the maximal and minimal domains $\Omega^{M_N^{n,l-1}}$ and $\Omega^{m_N^{n,l-1}}$ determined by the maximum and minimum functions $M_N^{n,l-1}$ and $m_N^{n,l-1}$ given by Lemma 2.5. We define the "common" test space required by the general property *(C1)* to be the space consisting of all the (smooth) functions $(\mathbf{q}, \psi)$ such that $\mathbf{q}$ is defined on $\Omega^{m_N^{n,l-1}}$ and then extended to the maximal domain $\Omega^{M_N^{n,l-1}}$ by the trace $\psi \mathbf{e}_r$ (which is constant in the $\mathbf{e}_r$ direction):

$$Q_N^{n,l} = \{(\mathbf{q}, \psi) \in \left(V_F(\Omega^{M_N^{n,l-1}}) \cap H^3(\Omega^{M_N^{n,l-1}})\right) \times H^3(\Gamma) : \mathbf{q} = \psi \mathbf{e}_r \text{ on } \Omega^{M_N^{n,l-1}} \setminus \Omega^{m_N^{n,l-1}}\}.$$

See Fig. 2.8.

We want to show that the space $Q_N^{n,l}$ is continuously embedded in all the test spaces Q_N^{n+i} for the time-shifts $i = 0, \dots, l$. Here, recall, that Q_N^{n+i} consists of the extensions by zero to the maximal domain of the functions defined in (2.6.8). In order to do that we construct the mapping $J_{N,l,n}^i$ which will provide the continuous embedding, and will satisfy the desired properties in *(C1)*. The mapping $J_{N,l,n}^i :$ $Q_N^{n,l} \to Q_N^{n+i}$ is defined by the following:

$$J_{N,l,n}^i(\mathbf{q}, \psi) = (\mathbf{q}_{N,l,n}^i, \psi),$$

where $(\mathbf{q}_{N,l,n}^i, \psi)$ are defined to be the **restrictions** to Ω_N^{n+i} of the functions in $Q_N^{n,l}$. See Fig. 2.8. More precisely,

$$\mathbf{q}_{N,l,n}^i = \begin{cases} \mathbf{q}, & \text{in } \Omega^{\eta_N^{n+i}} \\ 0, & \text{elsewhere} \end{cases}, \quad \mathbf{q} \in Q_N^{n,l}. \qquad (2.6.18)$$

From the definition of the test space $Q_N^{n,l}$ we can see that indeed, $J_{N,l,n}^i(\mathbf{q}, \psi) \in Q_N^{n+i}$, and that the functions from $Q_N^{n,l}$ are admissible for problem (2.6.5) for all domains $\Omega^{\eta_N^{n+i}}$. In particular, they are divergence-free for all i. Moreover,

$$\|J_{N,l,n}^i \mathbf{q}\|_{Q_N^{n+i}} \le C \|\mathbf{q}\|_{Q_N^{n,l}}.$$

To check that inequality (2.6.11) holds, we use Lemma 2.5 and compute:

$$\|J_{N,l,n}^{i}(\mathbf{q},\psi) - J_{N,l,n}^{j}(\mathbf{q},\psi)\|_{L^2(\Omega^M)} = \sqrt{\left|\int_0^L \int_{\eta_N^{n+i}(z)}^{\eta_N^{n+j}(z)} (\psi(z))^2\, dr\, dz\right|}$$

$$= \sqrt{\left|\int_0^L (\psi(z))^2\, (\eta_N^{n+j}(z) - \eta_N^{n+i}(z))\, dz\right|}$$

$$\leq \sqrt{\|\psi\|_{L^\infty}^2 \|\eta_N^{n+j} - \eta_N^{n+i}\|_{L^1(\Gamma)}} \leq C\|(\mathbf{q},\psi)\|_{Q_N^{n,l}} \sqrt{\|\eta_N^{n+j} - \eta_N^{n+i}\|_{L^1(\Gamma)}}.$$

We further estimate the right hand-side as follows:

$$\|\eta_N^{n+j} - \eta_N^{n+i}\|_{L^1(\Gamma)} = \int_\Gamma |\eta_N^{n+j} - \eta_N^{n+j-1} + \eta_N^{n+j-1} + \cdots - \eta_N^{n+i+1} + \eta_N^{n+i+1} - \eta_N^{n+i}|\, dz$$

$$\leq \int_\Gamma \sum_{k=i+1}^{j} |\eta_N^{n+k} - \eta_N^{n+k-1}|\, dz \leq C \sum_{k=i+1}^{j} \|\eta_N^{n+k} - \eta_N^{n+k-1}\|_{L^2(\Gamma)}$$

$$= C \sum_{k=i+1}^{j} \|v_N^{n+k}\|_{L^2(\Gamma)} \Delta t \leq C|j-i|\Delta t,$$

where in the last inequality we used the uniform energy bound for the structure velocities v_N^{n+k}, which were defined by $v_N^{n+k} = (\eta_N^{n+k} - \eta_N^{n+k-1})/\Delta t$. By combining these estimates one gets:

$$\|J_{N,l,n}^{i}(\mathbf{q},\psi) - J_{N,l,n}^{j}(\mathbf{q},\psi)\|_{L^2(\Omega)} \leq C\|(\mathbf{q},\psi)\|_{Q_N^{n,l}} \sqrt{|i-j|\Delta t}.$$

To check that inequality (2.6.12) holds, we recall that $J_{N,l,n}^{i}\mathbf{q}$ and $\mathbf{q}$ differ only in the region $\Omega^M \setminus \Omega^{\eta_N^{n+i}}$, and so the H-norm difference between the two functions can be bounded by the $Q_N^{n,l}$-norm of $\mathbf{q}$ and the L^2-norm of the difference between $M_N^{n,l}$ and $M_N^{n,l}$, which is bounded by $Cl\Delta t$ according to Lemma 2.5. This completes the construction of a common test space for the time shifts, as specified by property (C1).

Property C2: Approximation Property of Solution Spaces We would like to be able to compare and estimate the time shifts by h of the fluid velocity function $\mathbf{u}$, given at time $t+h$, with the function $\mathbf{u}$ given at time t. Again, as we mentioned earlier, the time-shift and the function itself are defined on different physical domains, since they also depend on time. We would like to define a common solution space on which we can make the function comparisons, where the common function space needs to be constructed in such a way that the functions from that space approximate well the original functions defined on different domains. The common space will be defined on a domain that is contained in all the other domains (we can do this because our approximate domains are close). The functions corresponding to the time-shifts are going to be defined on this common domain by a "squeezing" procedure, also used in [21], and then extended to the largest domain by the trace on the boundary in a way that keeps the divergence free condition satisfied. Keeping in mind that $h = l\Delta t$, we introduce the space $V_N^{n,l}$ to be the closure of $Q_N^{n,l}$ in V. We would like to define the mappings $I_{N,l,n}^i$:

$$I_{N,l,n}^i : V_N^{n+i} \to V_N^{n,l}, \ i = 0, 1, \dots, l, \ \text{where } V_N^{n,l} = \overline{Q_N^{n,l}}^V,$$

to approximate the functions in V_N^{n+i}, by the functions from $V_N^{n,l}$, which are all defined on the same domain. This will be done via a "squeezing" operator, defined below (see also [21]), and then extending the squeezed functions onto the larger, common domain, by the trace v on the boundary.

Definition 2.4 (The Squeezing Operator) Let η_m, η, η_M be three functions defined on $[0, L]$ such that $-R < \eta_m(z) \le \eta(z) \le \eta_M(z)$, $z \in [0, L]$, so that $R + \eta_m$ defines Ω^{η_m} with $R + \eta_m > 0$. Let $\mathbf{u}$ be a divergence-free function defined on Ω^η such that $\mathbf{u} = v\mathbf{e}_r$ on Γ^η. For a given $\sigma \ge 1$, such that $\sigma(R + \eta_m) \ge R + \eta$, we define $\mathbf{u}_\sigma \in H^1(\Omega^{\eta_M})$ in the following way:

$$\mathbf{u}_\sigma(z, r) = \begin{cases} (\sigma u_z(z, \sigma r), u_r(z, \sigma r)), & \sigma r \le R + \eta(z), \\ v\mathbf{e}_r, & \text{elsewhere.} \end{cases} \tag{2.6.19}$$

This "squeezing" operator associates to any function $\mathbf{u}$ defined on Ω^η a function $\mathbf{u}_\sigma$ defined on the large domain Ω^{η_M}, containing all the important information about the original function $\mathbf{u}$ squeezed within the minimal domain Ω^{η_m}. See Fig. 2.9. The operator is designed by: (1) first squeezing the function $\mathbf{u}$ from domain Ω^η into Ω^{η_m} and rescaling the function u_z by σ so that the divergence free condition remains

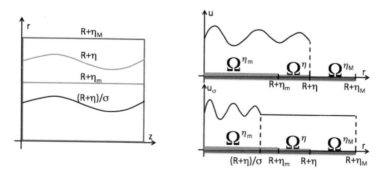

Fig. 2.9 The squeezing operator. Left: a sketch of the fluid domains. Right: a sketch an r-cut of $\mathbf{u}$ and $\mathbf{u}_\sigma$

to be satisfied, and (2) extending the squeezed $\mathbf{u}$ to the remainder of the maximal domain Ω^{η_M} by the values of its trace on η, where the extension is constant in the $\mathbf{e}_r$ direction.

Remark 2.11 Notice that $\nabla \cdot \mathbf{u} = 0$ implies that also $\nabla \cdot \mathbf{u}_\sigma = 0$. Moreover if $(\mathbf{u}, v) \in H^s(\Omega^\eta) \times H^s(0, L)$, then $\mathbf{u}_\sigma \in H^s(\Omega^{\eta_M})$, $0 \leq s \leq 1$.

We now define the mappings $I_{N,l,n}^i : V_N^{n+i} \to V_N^{n,l}$ by:

$$I_{N,l,n}^i : \mathbf{u}_N^{n+i} \mapsto (\mathbf{u}_N^{n+i})_{\sigma_{n,l}^i} \in V_N^{n,l}, \tag{2.6.20}$$

where $(\mathbf{u}_N^{n+i})_{\sigma_{n,l}^i}$ is obtained from Definition 2.4 by setting:

- $\eta_M = M_N^{n-1,l+1}$, $\eta = \eta_N^{n+i-1}$, $\eta_m = m_N^{n,l}$, and
- $\sigma_{n,l}^i > 0$ is such that $\sigma_{n,l}^i \geq \max\limits_{z \in [0,L]} \dfrac{R + M^{n-1,l+1}(z)}{R + \eta^{n+i-1}(z)}$.

Thus, the mappings $I_{N,l,n}^i$ take the functions $\mathbf{u}_N^{n+i}$ defined by the time-shits $i \Delta t$, $i = 0, \dots, l$, and associate to those functions the functions in the common velocity space $V_N^{n,l}$ that are all defined on the same domain $\Omega^{M_N^{n,l}}$, and are obtained by squeezing the functions into a domain which is contained in the minimal domain $\Omega^{m_N^{n,l}}$, and then extending them to $\Omega^{M_N^{n,l}}$ by their trace v_N^{n+1} on the moving boundary.

We need to prove that there exists a universal constant $C > 0$ such that

$$\|I_{N,l,n}^i(\mathbf{u}_N^{n+i})\|_{V_N^{n,l}} \leq C \|\mathbf{u}_N^{n+i}\|_{V_N^{n+i}}, \quad i = 0, \dots, l \tag{2.6.21}$$

and a universal, monotonically increasing function g, which converges to 0 as $h \to 0$, where $h = l\Delta t$, such that

$$\|I_{N,l,n}^i(\mathbf{u}_N^{n+i}) - \mathbf{u}_N^{n+i}\|_{L^2(\Omega^M)} \leq g(l\Delta t)\|\mathbf{u}_N^{n+i}\|_{V_N^{n+i}}, \quad i = 0, \dots, l. \tag{2.6.22}$$

The first inequality follows from Lemma 2.7 in [60] and the fact that the squeezed functions remain in H^1, with the uniformly bounded norm provided by the energy estimates in Lemma 2.1.

To show (2.6.22) we first prove the following lemma, which states that the difference between $\mathbf{u}$ and $\mathbf{u}_\sigma$ on Ω^M can be estimated by the L^2-norms of the gradient of $\mathbf{u}$ on Ω^η and the L^2-norm of the velocity on the boundary, where the constant in the estimate depends only on the Lipschitz constant of the fluid domain boundary.

Lemma 2.7 *Let η_m, η, and η_M be as in Definition 2.4, $\mathbf{u} \in H^1(\Omega^\eta)$, $\mathbf{u} = v\mathbf{e}_2$ on Γ^η, and let σ be such that $\sigma \geq \max\limits_{z \in [0,L]} \dfrac{R + \eta_M(z)}{R + \eta(z)}$. Then the following estimate holds:*

$$\|\mathbf{u} - \mathbf{u}_\sigma\|_{L^2(\Omega^{\eta_M})} \leq C\sqrt{\sigma - 1}(\|\nabla\mathbf{u}\|_{L^2(\Omega^\eta)} + \|v\|_{L^2(0,L)}),$$

where C depends only on $\|R + \eta(z))\|_{L^\infty}$.

Proof We will prove this estimate in two steps. First, we obtain an $L^2(\Omega^{\eta_M})$ estimate of the difference between $\mathbf{u}$ and a slightly modified function $\widetilde{\mathbf{u}}_\sigma$, which is *not divergence free*, defined by:

$$\widetilde{\mathbf{u}}_\sigma(z, r) = \begin{cases} (u_z(z, \sigma r), u_r(z, \sigma r)), & \sigma r \leq R + \eta(z), \\ v\mathbf{e}_r, & \text{elsewhere,} \end{cases}$$

and then estimate the difference between $\widetilde{\mathbf{u}}_\sigma$ and $\mathbf{u}_\sigma$. Notice that the only difference between $\widetilde{\mathbf{u}}_\sigma$ and $\mathbf{u}_\sigma$ is the factor σ in front of u_z.

The L^2-norm of the difference between $\mathbf{u}$ and $\widetilde{\mathbf{u}}_\sigma$ can be broken into three parts:

$$\int_{\Omega^{\eta_M}} |\mathbf{u}(z, r) - \widetilde{\mathbf{u}}_\sigma(z, r)|^2 dr dz = \int_0^L \int_0^{\frac{R+\eta(z)}{\sigma}} |\mathbf{u}(z, r) - \widetilde{\mathbf{u}}_\sigma(z, r)|^2 dr dz$$

$$+ \int_0^L \int_{\frac{R+\eta(z)}{\sigma}}^{R+\eta} |\mathbf{u}(z, r) - \widetilde{\mathbf{u}}_\sigma(z, r)|^2 dr dz + \int_0^L \int_{R+\eta}^{R+\eta_M} |\mathbf{u}(z, r) - \widetilde{\mathbf{u}}_\sigma(z, r)|^2 dr dz$$

$$= I_1 + I_2 + I_3.$$

The first part, I_1, can be estimated as follows:

$$I_1 = \int_0^L \int_0^{\frac{R+\eta(z)}{\sigma}} |\mathbf{u}(z, r) - \mathbf{u}(z, \sigma r)|^2 dr dz = \int_0^L \int_0^{\frac{R+\eta(z)}{\sigma}} |\partial_r \mathbf{u}(z, \xi)|^2 |r(1 - \sigma)|^2 dr dz$$

$$\leq C\|\nabla\mathbf{u}\|_{L^2(\Omega^{\eta_m})}^2 (1 - \sigma)^2.$$

Notice that the constant C depends on the size of Ω^{η_m}, which can be estimated from above by the size of the maximal domain.

To estimate I_2 we recall that $\widetilde{\mathbf{u}}_\sigma$ is defined to be $v\mathbf{e}_r$ outside the squeezed domain bounded by $(R + \eta(z))/\sigma$, which is exactly the trace of $\mathbf{u}$ on $R + \eta(z)$. Therefore, we get:

$$I_2 = \int_0^L \int_{\frac{R+\eta}{\sigma}}^{R+\eta} |\mathbf{u}(z, r) - \mathbf{u}(z, R + \eta(z))|^2 dr \, dz$$

$$\leq \int_0^L \int_{\frac{R+\eta}{\sigma}}^{R+\eta} |\partial_r \mathbf{u}(z, \xi)|^2 (R + \eta(z))^2 (1 - \frac{1}{\sigma})^2 \leq C(\sigma - 1)^2 \|\nabla\mathbf{u}\|_{L^2(\Omega^{\eta_m})}^2.$$

Finally, to estimate I_3 we recall that $\mathbf{u}$ is extended by zero outside Ω^η, and that, as before, $\widetilde{\mathbf{u}}_\sigma$ is equal to $v e_r$ in the same region. Therefore, we obtain:

$$I_3 = \int_0^L \int_{R+\eta}^{R+\eta_M} |v(z)|^2 dr dz \leq \int_0^L |v(z)|^2 (\eta_M - \eta) \leq \|v\|_{L^2(0,L)}^2 (\sigma - 1) \|R + \eta(z))\|_{L^\infty},$$

where the last inequality holds due to the assumption $\sigma \geq \max_{z \in [0,L]} \dfrac{R + \eta_M(z)}{R + \eta(z)}$.

In order to complete the proof and take into account the multiplication of u_z by σ, we use the triangle inequality:

$$\|u_z - (u_\sigma)_z\|_{L^2} \leq \|u_z - (u_\sigma)_z \pm (\widetilde{u}_\sigma)_z\|_{L^2} = \|u_z - (\widetilde{u}_\sigma)_z\|_{L^2} + \|u_z\|_{L^2}(\sigma - 1)$$

$$\leq C\sqrt{\sigma - 1}(\|\nabla \mathbf{u}\|_{L^2} + \|v\|_{L^2}).$$

This proves the lemma.

To show (2.6.22) we use this lemma to get:

$$\|I_{N,n,l}^i \mathbf{u}_N^{n+i} - \mathbf{u}_N^{n+i}\|_{L^2(\Omega^{\eta_M})} \leq C\sqrt{\sigma_{n,l}^i - 1}\|\mathbf{u}_N^{n+i}\|_{V_N^{n+i}} \leq C\sqrt{\sigma_{n,l} - 1}\|\mathbf{u}_N^{n+i}\|_{V_N^{n+i}},$$

where

$$\sigma_{n,l} := \max_{z \in [0,L]} \frac{R + M_N^{n-1,l+1}}{R + m_N^{n,l}}.$$

It is now crucial to show that we can estimate $(\sigma_{n,l} - 1)$ appearing on the right hand-side of the above inequality by a function $g(h)$ such that $g \to 0$ as $h \to 0$, where $h = l \Delta t$. Indeed, we have:

$$\sigma_{n,l} - 1 = \frac{R + M_N^{n-1,l+1}}{R + m_N^{n,l}} - \frac{R + m_N^{n,l}}{R + m_N^{n,l}} = \frac{M_N^{n-1,l+1} - m_N^{n,l}}{R + m_N^{n,l}}$$

$$\leq C(M_N^{n-1,l+1} - m_N^{n,l}) \leq C\sqrt{l \Delta t},$$

where the last inequality follows from Lemma 2.5. Thus, we have shown that $g(h) = Ch^{1/4}$, and that the following density results holds:

Corollary 2.1 (Density) *Let $n, l, N \in \mathbb{N}$ be such that $n + l \leq N$, $h = l \Delta t$ and $i \in \{0, \dots, l\}$. Let $(\mathbf{u}, v) \in V_N^{n+i}$. Then there exists a constant C (independent of N, n, l) and the operators $I_{N,n,l}^i$ defined by (2.6.20) such that $I_{N,n,l}^i(\mathbf{u}, v) \in V_N^{n,l}$ and*

$$\|I_{N,n,l}^i \mathbf{u} - \mathbf{u}\|_{L^2} \leq Ch^{1/4}\|\mathbf{u}_N^{n+i}\|_{V_N^{n+i}}, \quad i \in \{0, \dots, l\}.$$

Property C3: Uniform Ehrling Property We need to prove the Uniform Ehrling Property, stated in (2.6.15). The main difficulty comes, again, from the fact that we have to work with moving domains, which are parameterized by N, n and l.

Remark 2.12 (Notation) To show that the uniform Ehrling estimate (2.6.15) holds, independently of all three parameters, we simplify notation (only in this proof) and replace the indices N, n, l with only one index, n, so that, e.g., the minimum and maximum functions $m_N^{n,l}$, $M_N^{n,l}$ are re-enumerated as M_n, m_n, with the corresponding maximal fluid domains Ω^{M_n} and the function spaces H_n, V_n and Q_n' all defined on Ω^{M_n}.

We prove the Uniform Ehrling Property by contradiction, and by using the "uniform squeezing property" ([68], Lemma 5.7). We start by assuming that the statement of the uniform Ehrling property (2.6.15) is false. More precisely, we assume that there exists a $\delta_0 > 0$ and a sequence $(\mathbf{f}_n, g_n) \in H_n$ such that

$$\|(\mathbf{f}_n, g_n)\|_H = \|(\mathbf{f}_n, g_n)\|_{H_n} > \delta_0 \|(\mathbf{f}_n, g_n)\|_{V_n} + n\|(\mathbf{f}_n, g_n)\|_{Q_n'}.$$

Here, as before, we have extended the functions f_n onto the entire domain Ω^M, which is determined by the maximal function $M(z)$ defined in Lemma 2.5, so that $\|(\mathbf{f}_n, g_n)\|_H = \|(\mathbf{f}_n, g_n)\|_{H_n}$. Recall that H is defined as the L^2 product space, so extensions by 0 do not change the norm.

It will be convenient to also replace the V_n norm on the right hand-side by the norm on V. Here, however, since V is defined as the H^s product space, the norm on V_n is bounded from below by a constant times the norm on V, see Lemma 2.6. Thus, we have:

$$\|(\mathbf{f}_n, g_n)\|_H > \delta_0 \|(\mathbf{f}_n, g_n)\|_{V_n} + n\|(\mathbf{f}_n, g_n)\|_{Q_n'} \geq C\delta_0 \|(\mathbf{f}_n, g_n)\|_V + n\|(\mathbf{f}_n, g_n)\|_{Q_n'}.$$

Without the loss of generality we can assume that our sequence $(\mathbf{f}_n, g_n)$ is such that $\|(\mathbf{f}_n, g_n)\|_H = 1$. For example, we could consider $\frac{1}{\|(\mathbf{f}_n, g_n)\|_H}(\mathbf{f}_n, g_n)$ instead of $(\mathbf{f}_n, g_n)$. Notice that we now have a sequence such that the two terms on the right hand-side are uniformly bounded in n, which implies that there exists a subsequence, which we again denote by $(\mathbf{f}_n, g_n)$, such that:

$$\|(\mathbf{f}_n, g_n)\|_H = 1, \quad \|(\mathbf{f}_n, g_n)\|_V \leq \frac{1}{C\delta_0}, \quad \|(\mathbf{f}_n, g_n)\|_{Q_n'} \to 0. \qquad (2.6.23)$$

Since $(\mathbf{f}_n, g_n)$ is uniformly bounded in V, and by the compactness of the embedding of V into H, we conclude that there exists a subsequence $(\mathbf{f}_{n_k}, g_{n_k}) \to (\mathbf{f}, g)$ strongly in H.

Now, from the energy estimates, we recall that sequences $\{M_n\}$, $\{m_n\}$ are uniformly bounded in $H^2(\Gamma)$, and so by the compactness of the embedding of $H^2(\Gamma)$ into $C(\bar{\Gamma})$ there exist subsequences $M_{n_k} \to \widetilde{M}$, $m_{n_k} \to \widetilde{m}$. With a slight abuse of notation, we denote the convergent subsequences by index n. Therefore, we are now working with the convergent sub-sequences

$$M_n \to \widetilde{M},\ m_n \to \widetilde{m},\ (\mathbf{f}_n, g_n) \to (\mathbf{f}, g) \text{ in } H.$$

These functions define the maximal domain $\Omega^{\widetilde{M}}$, the minimal domain $\Omega^{\widetilde{m}}$, and the corresponding minimal and maximal domains that depend on n: Ω^{m_n} and Ω^{M_n}.

Let $\widetilde{S}$ be the strip between the minimal and maximal domains $\Omega^{\widetilde{m}}$ and $\Omega^{\widetilde{M}}$, and S_n be the strip between the minimal and maximal domains that depend on n:

$$S_n = \Omega^{M_n} \setminus \Omega^{m_n}, \quad \widetilde{S} = \Omega^{\widetilde{M}} \setminus \Omega^{\widetilde{m}}.$$

We now want to show that $(\mathbf{f}, g) = 0$ first outside $\Omega^{\widetilde{M}}$, i.e., in $\Omega^M \setminus \Omega^{\widetilde{M}}$, then in $\Omega^{\widetilde{M}}$. (Recall that Ω^M was defined to contain all the approximate domains.) This will contradict the assumption that $\|(\mathbf{f}_n, g_n)\|_H = 1$ and complete the proof.

First, we show that $\mathbf{f} = 0$ outside of $\Omega^{\widetilde{M}}$, and inside Ω^M. This will follow simply because $\mathbf{f}$ is the limit of a sequence of functions that are zero outside Ω^{M_n} which converge to $\Omega^{\widetilde{M}}$. More precisely, we introduce the characteristic functions $\chi_{\Omega^{M_n}}$ of the sets Ω^{M_n} and recall that $\chi_{\Omega^{M_n}}$ converge uniformly to $\chi_{\Omega^{\widetilde{M}}}$. Therefore:

$$\mathbf{f}(1 - \chi_{\Omega^{\widetilde{M}}}) = \lim_n \mathbf{f}_n(1 - \chi_{\Omega^{M_n}}) = 0,$$

since $\mathbf{f}_n$ are extended by 0 to Ω^M.

Next, we show that $\mathbf{f} = 0$ inside $\Omega^{\widetilde{M}}$. We start by showing that $\mathbf{f} = g\mathbf{e}_r$ in $\widetilde{S}$. This follows immediately from

$$(\mathbf{f} - g\mathbf{e}_r)\chi_{\widetilde{S}} = \lim_n (\mathbf{f}_n - g_n\mathbf{e}_r)\chi_{S_n} = 0,$$

because $(\mathbf{f}_n, g_n) \in V_n = \overline{Q_n}^V$, which implies $\mathbf{f}_n = g_n\mathbf{e}_r$ in S_n.

We finish the proof by using $\|(\mathbf{f}_n, g_n)\|_{Q'_n} \to 0$ to show $(\mathbf{f}, g) = (\mathbf{0}, 0)$. Recall that:

$$Q_n = \{(\mathbf{q}, \psi) \in \left(V_F(\Omega^{M_n}) \cap H^4(\Omega^{M_n})\right) \times H_0^2(\Gamma) : \mathbf{q} = \psi\mathbf{e}_3 \text{ on } S_n\}.$$

Let $\varepsilon > 0$. By the density of Q_n in H, and by the uniform convergence of M_n and m_n, combined with the "squeezing procedure" from the proof of Property C2, we can construct a test function $(\mathbf{q}, \psi)$ and $n_0 \in \mathbb{N}$ such that:

$$\|(\mathbf{f}, g) - (\mathbf{q}, \psi)\|_H \leq \varepsilon, \quad (\mathbf{q}, \psi) \in Q_n,\ n \geq n_0. \tag{2.6.24}$$

Therefore we have:

$$\langle (\mathbf{f}, g), (\mathbf{q}, \psi) \rangle_H = \lim_n \langle (\mathbf{f}_n, g_n), (\mathbf{q}, \psi) \rangle_H = \lim_n {}_{Q'_n} \langle (\mathbf{f}_n, g_n), (\mathbf{q}, \psi) \rangle_{Q_n}$$

$$\leq \| (\mathbf{f}_n, g_n) \|_{Q'_n} \underbrace{\| (\mathbf{q}, \psi) \|_{Q_n}}_{\leq C} \to 0.$$

Here, the uniform boundedness of $\| (\mathbf{q}, \psi) \|_{Q_n}$ follows from the uniform convergence of M_n and m_n. Hence,

$$\| (\mathbf{f}, g) \|_H^2 = \langle (\mathbf{f}, g), (\mathbf{f}, g) \pm (\mathbf{q}, \psi) \rangle_H = \langle (\mathbf{f}, g), (\mathbf{f} - \mathbf{q}, g - \psi) \rangle_H \leq \varepsilon \| (\mathbf{f}, g) \|_H.$$

Therefore, $\| (\mathbf{f}, g) \|_H < \varepsilon$. Since ε is arbitrary, $(\mathbf{f}, g) = (\mathbf{0}, 0)$, and this completes the proof of the Uniform Ehrling Property.

Since this was the last step in verifying that all the assumptions of Theorem 2.1 hold, we conclude that the statement of the compactness theorem, Theorem 2.2, holds true.

We summarize the strong convergence results obtained in this section. We have shown that there exist subsequences $(\mathbf{u}_N)_{N \in \mathbb{N}}$, $(\eta_N)_{N \in \mathbb{N}}$, $(\bar{\eta}_N)_{N \in \mathbb{N}}$, $(v_N)_{N \in \mathbb{N}}$ such that

$$\mathbf{u}_N \to \mathbf{u} \text{ in } L^2(0, T; L^2(\Omega_M)),$$

$$\tau_{\Delta t} \mathbf{u}_N \to \mathbf{u} \text{ in } L^2(0, T; L^2(\Omega_M)),$$

$$\eta_N \to \eta \text{ in } L^\infty(0, T; C^1(\bar{\Gamma})),$$

$$\bar{\eta}_N \to \bar{\eta} \text{ in } L^\infty(0, T; C^1(\bar{\Gamma})), \tag{2.6.25}$$

$$v_N \to v \text{ in } L^2(0, T; L^2(\Gamma)),$$

$$\tau_{\Delta t} v_N \to v \text{ in } L^2(0, T; L^2(\Gamma)).$$

The statements about the convergence of the time-shifts $(\tau_{\Delta t} \mathbf{u}_N)_{N \in \mathbb{N}}$ and $(\tau_{\Delta t} v_N)_{N \in \mathbb{N}}$ follow directly from Statement 3 of Theorem 2.1, i.e., from the numerical dissipation estimate for the splitting scheme. Namely, if we multiply the third equality of Lemma 2.1 by Δt we get:

$$\| \tau_{\Delta t} \mathbf{u}_N - \mathbf{u}_N \|^2_{L^2((0,T) \times \Omega)} + \| \tau_{\Delta t} v_N - v_N \|^2_{L^2((0,T) \times (0,L))} \leq C \Delta t. \tag{2.6.26}$$

Furthermore, one can also show that subsequences $(\tilde{v}_N)_N$ and $(\tilde{\mathbf{u}}_N)_N$ also converge to v and $\mathbf{u}$, respectively. More precisely,

$$\tilde{\mathbf{u}}_N \to \mathbf{u} \text{ in } L^2(0, T; L^2(\Omega)),$$
$$\tilde{v}_N \to v \text{ in } L^2(0, T; L^2(0, L)). \tag{2.6.27}$$

This statement follows directly from the following inequalities (see [75, p. 328])

$$\| v_N - \widetilde{v}_N \|^2_{L^2(0,T;L^2(0,L))} \leq \frac{\Delta t}{3} \sum_{n=1}^N \| v^{n+1} - v^n \|^2_{L^2(0,L)},$$

$$\| \mathbf{u}_N - \widetilde{\mathbf{u}}_N \|^2_{L^2(0,T;L^2(\Omega))} \leq \frac{\Delta t}{3} \sum_{n=1}^N \| \mathbf{u}^{n+1} - \mathbf{u}^n \|^2_{L^2(\Omega)},$$

and Lemma 2.1 which provides uniform boundedness of the sums on the right hand-sides of the inequalities.

2.7 The Limiting Problem and Existence of Weak Solutions

One of the problems in showing that the sequence of approximate solutions converges to a weak solution of the coupled, continuous problem, is the fact that the velocity test functions depend on the fluid domains, i.e., they depend on N (Δt). We would ultimately like to show that in the limit as $N \to \infty$, the limits of approximate solutions satisfy the weak formulation stated in Sect. 2.3, for all the test functions in the given test space specified in Sect. 2.3. To take the limit in the weak formulations of approximate problems, we need to be able to "handle" the behavior of test functions as $N \to \infty$. In particular, we would like to show that not only the approximate solutions, but also the corresponding velocity test functions converge to the velocity test functions of the continuous problem. Moreover, the convergence needs to be strong enough (uniform) to allow passing to the limit in the integral formulations.

To achieve this goal, we start by considering the velocity test functions $\mathbf{q}$ for the continuous, coupled problem, and based on those functions "construct" the test functions $\mathbf{q}_N$ for approximate problems, such that they have the following two properties: (1) they belong to the test spaces of approximate problems, and (2) they converge (uniformly), as $N \to \infty$, to a test function $\mathbf{q}$ of the continuous problem. There are different ways to construct such test functions. Here, we take an approach similar to [21].

2.7.1 Construction of the Appropriate Test Functions

Recall that the test functions $(\mathbf{q}, \psi)$ for the limiting problem are defined by the space Q^η, given in (2.3.7), which depends on η. Similarly, the test spaces for the approximate problems depend on N through the dependence on η_N.

Fig. 2.10 Construction of appropriate test functions $\mathbf{q}_0$ and $\mathbf{q}_1$ for the limiting procedure

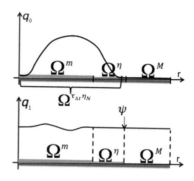

We begin the construction of the appropriate test functions by considering a dense subset $X \subset Q^\eta$ such that X consists of all the test functions $(\mathbf{q}, \psi) \in X = X_F \times X_S$, where X_F is such that the velocity test functions $\mathbf{q} \in X_F$ are smooth, independent of N, and $\nabla \cdot \mathbf{q} = 0$.

To construct the set X_F we consider the functions $(\mathbf{q}, \psi) \in Q^\eta$ which can be written as an algebraic sum of the functions $\mathbf{q}_0$, which have compact support in $\Omega^\eta \cup \Gamma_{in} \cup \Gamma_{out} \cup \Gamma_b$, plus a function $\mathbf{q}_1$, which is a "constant extension" of ψ (the Lagrangian trace of $\mathbf{q}$):

$$\mathbf{q} = \mathbf{q}_0 + \mathbf{q}_1.$$

More precisely, the functions $\mathbf{q}_0$ and $\mathbf{q}_1$ can be constructed as follows. Let Ω_{min} and Ω_{max} denote the fluid domains associated with the radii R_{min} and R_{max}, respectively. See Fig. 2.10.

1. **Definition of test functions $(\mathbf{q}_0, 0)$ on $(0, T) \times \Omega_{max}$:** Consider all the smooth functions $\mathbf{q}$ with compact support in $\Omega^\eta \cup \Gamma_{in} \cup \Gamma_{out} \cup \Gamma_b$, and such that $\nabla \cdot \mathbf{q} = 0$. Then we can extend $\mathbf{q}$ by 0 to a divergence-free vector field on $(0, T) \times \Omega_{max}$. This defines $\mathbf{q}_0$.

2. **Definition of test functions $(\mathbf{q}_1, \psi)$ on $(0, T) \times \Omega_{max}$:** Consider $\psi \in C_c^1([0, T); H_0^2(\Gamma^\eta))$. Define

$$\mathbf{q}_1 := \begin{cases} \begin{array}{l} \text{A constant extension in the vertical} \\ \text{direction of } \psi \mathbf{e}_r \text{ on } \Gamma^\eta : \mathbf{q}_1 = (0, \psi(z))^T; \\ \text{Notice, } \mathrm{div}\,\mathbf{q}_1 = 0. \end{array} \bigg\} \text{ on } \Omega_{max} \setminus \Omega_{min}, \\[1em] \text{A divergence-free extension to } \Omega_{min} \big\} \text{ on } \Omega_{min}. \end{cases}$$

See, e.g., [40, pg. 127].

Notice that since η_N converge uniformly to η, there exists an $N_q > 0$ such that $\mathrm{supp}(\mathbf{q}_0) \subset \Omega^{\tau_{\Delta t}\eta_N}$, $\forall N \geq N_q$. Therefore, $\mathbf{q}_0$ is well defined on infinitely many approximate domains $\Omega^{\tau_{\Delta t}\eta_N}$.

From the construction it is clear that $\mathbf{q}_1$ is also defined on $\Omega_{\tau_{\Delta t}\eta_N}$ for each N, and so it can be mapped onto the reference domain Ω by the transformation $\mathcal{A}^{\tau_{\Delta t}\eta_N}$.

For any test function $(\mathbf{q}, \psi) \in Q^\eta$ it is easy to see that the velocity component $\mathbf{q}$ can then be written as $\mathbf{q} = \mathbf{q} - \mathbf{q}_1 + \mathbf{q}_1$, where $\mathbf{q} - \mathbf{q}_1$ can be approximated by a divergence-free function $\mathbf{q}_0$, which has compact support in $\Omega^\eta \cup \Gamma_{in} \cup \Gamma_{out} \cup \Gamma_b$. Therefore, one can easily see that functions

$$(\mathbf{q}, \psi) = (\mathbf{q}_0 + \mathbf{q}_1, \psi) \in \mathcal{X}_F \times \mathcal{X}_S = \mathcal{X}$$

satisfy the following properties:

1. $\mathcal{X}$ is dense in the space Q^η of all test functions defined on the physical, moving domain Ω^η, defined by (2.3.7);
2. $\nabla \cdot \mathbf{q} = 0, \forall \mathbf{q} \in \mathcal{X}_F$;
3. For each $\mathbf{q} \in \mathcal{X}_F$, there exists an $N_q > 0$ such that $\forall N \geq N_q$, $\mathbf{q}_0$ and $\mathbf{q}_1$ are well defined on $\Omega^{\tau_{\Delta t}\eta_N}$.

Now, we can use those test functions to study convergence as $N \to \infty$ of the approximate solutions, defined on domains $\Omega^{\tau_{\Delta t}\eta_N}$. To do this, we map everything onto a fixed, reference domain via the ALE mappings $\mathcal{A}^{\tau_{\Delta t}\eta_N}$, and work on the reference domain Ω.

For this purpose, we define the test functions $\widetilde{\mathbf{q}}$ defined on Ω, to be the test functions $\mathbf{q} \in \mathcal{X}_F$ composed with the ALE mapping $\mathcal{A}^\eta$:

$$\widetilde{\mathbf{q}} = \mathbf{q} \circ \mathcal{A}^\eta.$$

The set $\{(\widetilde{\mathbf{q}}, \psi) | \widetilde{\mathbf{q}} = \mathbf{q} \circ \mathcal{A}^\eta, \mathbf{q} \in \mathcal{X}_F, \psi \in \mathcal{X}_S\}$ is dense in the space $\widetilde{Q}^\eta$ of all test functions defined on the fixed, reference domain Ω, defined by (2.3.11).

Similarly, for $\mathbf{q} \in \mathcal{X}_F$, we define $\widetilde{\mathbf{q}}_N$ defined on Ω, to be the test functions $\mathbf{q} \in \mathcal{X}_F$ composed with the ALE mapping $\mathcal{A}^{\tau_{\Delta t}\eta_N}$:

$$\widetilde{\mathbf{q}}_N := \mathbf{q} \circ \mathcal{A}^{\tau_{\Delta t}\eta_N}.$$

Functions $\widetilde{\mathbf{q}}_N$ satisfy $\nabla^{\tau_{\Delta t}\eta_N} \cdot \widetilde{\mathbf{q}}_N = 0$. These will serve as test functions for approximate problems, defined on domains determined by $\tau_{\Delta t}\eta_N$. The approximate test functions have the following uniform convergence properties, which are a consequence of uniform convergence of η_N and the spatial derivatives of η_N.

Lemma 2.8 *For every* $(\mathbf{q}, \psi) \in \mathcal{X}$ *we have*

$$\widetilde{\mathbf{q}}_N \to \widetilde{\mathbf{q}} \text{ and } \nabla\widetilde{\mathbf{q}}_N \to \nabla\widetilde{\mathbf{q}} \text{ uniformly on } [0, T] \times \Omega.$$

Proof By the Mean-Value Theorem we get:

$$|\widetilde{\mathbf{q}}_N(t, z, r) - \widetilde{\mathbf{q}}(t, z, r)| = |\mathbf{q}(t, z, (R + \tau_{\Delta t}\eta_N)r) - \mathbf{q}(t, z, (R + \eta)r)|$$
$$= |\partial_r \mathbf{q}(t, z, \zeta)r| \, |\eta(t, z) - \eta_N(t - \Delta t, z)|.$$

The uniform convergence of $\widetilde{\mathbf{q}}_N$ follows from the uniform convergence of η_N, since $\mathbf{q}$ are smooth.

To show the uniform convergence of the gradients, one can use the chain rule to calculate

$$\partial_z \widetilde{\mathbf{q}}_N(t, z, r) = \partial_z \mathbf{q}(t, z, (R + \tau_{\Delta t}\eta_N)r) + \left[\partial_z \tau_{\Delta t}\eta_N(t, z)r\right] \left[\partial_r \mathbf{q}(t, z, (R + \tau_{\Delta t}\eta_N)r)\right].$$

The uniform convergence of $\partial_z \widetilde{\mathbf{q}}_N(t, z, r)$ follows from the uniform convergence of $\partial_z \tau_{\Delta t}\eta_N$. Combined with the first part of the proof we get $\partial_z \widetilde{\mathbf{q}}_N \to \partial_z \widetilde{\mathbf{q}}$ uniformly on $[0, T] \times \Omega$. The uniform convergence of $\partial_r \widetilde{\mathbf{q}}_N$ can be shown in a similar way.

Before we can pass to the limit in the weak formulation of the approximate problems, there is one more useful observation that we need. Namely, notice that although $\mathbf{q}$ are smooth functions both in the spatial variables and in time, the functions $\widetilde{\mathbf{q}}_N$ are discontinuous at $n\Delta t$ because $\tau_{\Delta t}\eta_N$ is a step function in time. As we shall see below, it will be useful to approximate each discontinuous function $\widetilde{\mathbf{q}}_N$ in time by a *piecewise linear function*, $\bar{\mathbf{q}}_N$, so that

$$\bar{\mathbf{q}}_N(t, .) = \widetilde{\mathbf{q}}_N(n\Delta t-, .), \quad t \in [(n - 1)\Delta t, n\Delta t), \; n = 1, \ldots, N,$$

where $\widetilde{\mathbf{q}}_N(n\Delta t-)$ is the limit from the left of $\widetilde{\mathbf{q}}_N$ at $n\Delta t$, $n = 1, \ldots, N$. By combining Lemma 2.8 with the argument in the proof of Lemma 2.3, we get

$$\bar{\mathbf{q}}_N \to \widetilde{\mathbf{q}} \text{ uniformly on } [0, T] \times \Omega.$$

Remark 2.13 Notice that the "corresponding" test functions ψ_N are independent of N on Γ, and are equal to ψ there. The particular "constant" extension of ψ is crucial in avoiding the dependence of ψ_N on the regularity of η_N (via the ALE mapping).

2.7.2   Passing to the Limit

Notation Since in this section we will be only working with the problem defined on the fixed domain Ω, we will simplify notation for the test functions by using $\mathbf{q}_N$ to denote the functions $\mathbf{q} \circ \mathcal{A}^{\tau_{\Delta t}\eta_N} = \widetilde{\mathbf{q}}_N$ from above.

We start by writing the weak formulation of the coupled, semi-discretized problem (2.4.15) by using the test functions $(\mathbf{q}_N(t), \psi(t)) \in X$ (where $\mathbf{q}_N = \mathbf{q} \circ \mathcal{A}^{\tau_{\Delta t}\eta_N}$, $\mathbf{q} \in X_F$ and $\psi(t) \in X_S$). We integrate with respect to t from $n\Delta t$ to $(n+1)\Delta t$ and then take the sum from $n = 0, \ldots, N-1$ to get the time integrals over $(0, T)$. Furthermore, we use the fact that the following is true:

$$\rho_f \sum_{n=0}^{N-1} \int_\Omega (R + \tau_{\Delta t}\, \eta_N)\frac{\mathbf{u}_N^{n+1} - \mathbf{u}_N^n}{\Delta t} \cdot \mathbf{q}_N = \rho_f \int_0^T \int_\Omega (R + \tau_{\Delta t}\, \eta_N)\partial_t \bar{\mathbf{u}}_N \cdot \mathbf{q}_N$$

where $\bar{\mathbf{u}}_N$ is the *piecewise linear extrapolation*, as given by Definition 2.3. Now the weak formulation for a fixed N, i.e., fixed Δt, is given by the following:

$$\rho_f \int_0^T \int_\Omega (R + \tau_{\Delta t}\, \eta_N)\Big(\partial_t \bar{\mathbf{u}}_N \cdot \mathbf{q}_N + \frac{1}{2}(\tau_{\Delta t}\mathbf{u}_N - \mathbf{w}_N) \cdot \nabla^{\tau_{\Delta t}\eta_N}\mathbf{u}_N \cdot \mathbf{q}_N$$
$$-\frac{1}{2}(\tau_{\Delta t}\mathbf{u}_N - \mathbf{w}_N) \cdot \nabla^{\tau_{\Delta t}\eta_N}\mathbf{q}_N \cdot \mathbf{u}_N\Big) + \frac{\rho_f}{2}\int_0^T \int_\Omega v_N^* \mathbf{u}_N \cdot \mathbf{q}_N$$
$$+2\mu \int_0^T \int_\Omega (R + \tau_{\Delta t}\eta_N)\mathbf{D}^{\tau_{\Delta t}\eta_N}(\mathbf{u}_N) : \mathbf{D}^{\tau_{\Delta t}\eta_N}(\mathbf{q}_N) + \rho_s h \int_0^T \int_0^L \partial_t \bar{v}_N \psi$$
$$+ \int_0^T a_S(\eta_N, \psi)$$
$$= R\Big(\int_0^T P_{in}^N dt \int_0^R q_z(t, 0, r)dr - \int_0^T P_{out}^N dt \int_0^R q_z(t, L, r)dr\Big),$$
$$(2.7.1)$$

with

$$\nabla^{\tau_{\Delta t}\eta} \cdot \mathbf{u}_N = 0, \quad v_N = ((u_r)_N)|_\Gamma,$$
$$(2.7.2)$$
$$\mathbf{u}_N(0, .) = \mathbf{u}_0, \quad \eta(0, .)_N = \eta_0, \quad v_N(0, .) = v_0.$$

Here $\bar{\mathbf{u}}_N$ and $\bar{v}_N$ are the piecewise linear functions defined in Definition 2.3, $\tau_{\Delta t}$ is the shift in time by Δt to the left, defined in (2.6.10), $\nabla^{\tau_{\Delta t}\eta_N}$ is the transformed gradient via the ALE mapping $\mathcal{A}^{\tau_{\Delta t}\eta_N}$, defined in (2.2.29), and v_N^*, $\mathbf{u}_N$, v_N and η_N are defined in (2.5.1).

Using the convergence results obtained for the approximate functions in Sect. 2.6.2, and the convergence results just obtained for the test functions $\mathbf{q}_N$, we can pass to the limit directly in all the terms except in the term that contains $\partial_t \bar{\mathbf{u}}_N$.

To deal with this term we notice that, since $\mathbf{q}_N$ are smooth on sub-intervals $(n\Delta t, (n+1)\Delta t)$, we can use integration by parts on these sub-intervals to obtain:

$$\int_0^T \int_\Omega (R + \tau_{\Delta t} \eta_N) \partial_t \bar{\mathbf{u}}_N \cdot \mathbf{q}_N = \sum_{j=0}^{N-1} \int_{n\Delta t}^{(n+1)\Delta t} \int_\Omega (R + \eta_N^n) \partial_t \bar{\mathbf{u}}_N \cdot \mathbf{q}_N$$

$$= \sum_{j=0}^{N-1} \left(- \int_{n\Delta t}^{(n+1)\Delta t} \int_\Omega (R + \tau_{\Delta t} \eta_N) \bar{\mathbf{u}}_N \cdot \partial_t \mathbf{q}_N \right.$$

$$\left. + \int_\Omega (R + \eta_N^{n+1} - \eta_N^{n+1} + \eta_N^n) \mathbf{u}_N^{n+1} \cdot \mathbf{q}_N((n+1)\Delta t-) - \int_\Omega (R + \eta_N^n) \mathbf{u}_N^j \cdot \mathbf{q}_N(n\Delta t+) \right).$$

$$(2.7.3)$$

Here, we have denoted by $\mathbf{q}_N((n+1)\Delta t-)$ and $\mathbf{q}_N(n\Delta t+)$ the limits from the left and right, respectively, of $\mathbf{q}_N$ at the appropriate points.

The integral involving $\partial_t \mathbf{q}_N$ can be simplified by recalling that $\mathbf{q}_N(t, \widetilde{z}, \widetilde{r}) = \mathbf{q}(t, \mathcal{A}^{\eta_N}(t)(\widetilde{z}, \widetilde{r})) = \mathbf{q}(t, \widetilde{z}, (R + \eta_N(t, \widetilde{z}, \widetilde{r})))$, where η_N are constant on each sub-interval $(n\Delta t, (n+1)\Delta t)$. Thus, by the chain rule, we see that $\partial_t \mathbf{q}_N = \partial_t \mathbf{q}$ on $(n\Delta t, (n+1)\Delta t)$. After summing over all $j = 0, \ldots, N-1$ we obtain

$$- \sum_{j=0}^{N-1} \int_{n\Delta t}^{(n+1)\Delta t} \int_\Omega (R + \tau_{\Delta t} \eta_N) \bar{\mathbf{u}}_N \cdot \partial_t \mathbf{q}_N = - \int_0^T \int_\Omega (R + \tau_{\Delta t} \eta_N) \bar{\mathbf{u}}_N \cdot \partial_t \mathbf{q}.$$

To deal with the last two terms in (2.7.3) we calculate

$$\sum_{j=0}^{N-1} \left(\int_\Omega (R + \eta_N^{n+1} - \eta_N^{n+1} + \eta_N^n) \mathbf{u}_N^{n+1} \cdot \mathbf{q}_N((n+1)\Delta t-) - \int_\Omega (R + \eta_N^n) \mathbf{u}_N^n \cdot \mathbf{q}_N(n\Delta t+) \right)$$

$$= \sum_{j=0}^{N-1} \int_\Omega \left((R + \eta_N^{n+1}) \mathbf{u}_N^{n+1} \cdot \mathbf{q}_N((n+1)\Delta t-) - (\eta_N^{n+1} - \eta_N^n) \mathbf{u}_N^{n+1} \cdot \mathbf{q}_N((n+1)\Delta t-) \right)$$

$$- \int_\Omega (R + \eta_0) \mathbf{u}_0 \cdot \mathbf{q}(0) - \sum_{j=1}^{N-1} \int_\Omega (R + \eta_N^n) \mathbf{u}_N^n \cdot \mathbf{q}_N(n\Delta t+) \right) \Delta t$$

Now, we can write $(\eta^{n+1} - \eta^n)$ as $v^{j+\frac{1}{2}}\Delta t$, and rewrite the summation indices in the first term to obtain that the above expression is equal to

$$
= \sum_{j=1}^{N} \int_{\Omega} (R + \eta_N^n)\mathbf{u}_N^n \cdot \mathbf{q}_N(n\Delta t-) - \int_0^T \int_{\Omega} v_N^* \mathbf{u}_N \cdot \bar{\mathbf{q}}_N
$$
$$
- \int_{\Omega} (R + \eta_0)\mathbf{u}_0 \cdot \mathbf{q}(0) - \sum_{j=1}^{N-1} \int_{\Omega} (R + \eta_N^n)\mathbf{u}_N^n \cdot \mathbf{q}_N(n\Delta t+).
$$

Since the test functions have compact support in $[0, T)$, the value of the first term at $j = N$ is zero, and so we can combine the two sums to obtain that the above expression is equal to:

$$
= \sum_{j=1}^{N} \int_{\Omega} (R + \eta_N^n)\mathbf{u}_N^n \cdot (\mathbf{q}_N(n\Delta t-) - \mathbf{q}_N(n\Delta t+))
$$
$$
- \int_{\Omega} (R + \eta_0)\mathbf{u}_0 \cdot \mathbf{q}(0) - \int_0^T \int_{\Omega} v_N^* \mathbf{u}_N \cdot \bar{\mathbf{q}}_N.
$$

Now we know how to pass to the limit in all the terms expect the first one. We continue to rewrite the first expression by using the Mean Value Theorem to obtain:

$$
\mathbf{q}_N(n\Delta t-, z, r) - \mathbf{q}_N(n\Delta t+, z, r) = \mathbf{q}(n\Delta t, z, (R + \eta_N^n)r) - \mathbf{q}(n\Delta t, z, (R + \eta_N^{n+1})r) =
$$
$$
= \partial_r \mathbf{q}(n\Delta t, z, \zeta)r(\eta_N^n - \eta_N^{n+1}) = -\Delta t \partial_r \mathbf{q}(n\Delta t, z, \zeta)v_N^{j+\frac{1}{2}}r.
$$

Therefore we have:

$$
\sum_{j=1}^{N-1} \int_{\Omega} (R + \eta_N^n)\mathbf{u}_N^n(\mathbf{q}(n\Delta t-) - \mathbf{q}(n\Delta t+)) = -\int_0^{T-\Delta t} \int_{\Omega} (R + \eta_N)\mathbf{u}_N r \tau_{-\Delta t} v_N^* \partial_r \bar{\mathbf{q}}.
$$

We can now pass to the limit in this last term to obtain:

$$
\int_0^{T-\Delta t} \int_{\Omega} (R + \eta_N)\mathbf{u}_N r \tau_{-\Delta t} v_N^* \partial_r \bar{\mathbf{q}} \to \int_0^T \int_{\Omega} (R + \eta)\mathbf{u} r \partial_t \eta \partial_r \mathbf{q}.
$$

Therefore, by noticing that $\partial_t \bar{\mathbf{q}} = \partial_t \mathbf{q} + r \partial_t \eta \partial_r \mathbf{q}$ we have finally obtained

$$\int_0^T \int_\Omega (R + \tau_{\Delta t} \eta_N) \partial_t \bar{\mathbf{u}}_N \cdot \mathbf{q}_N \to - \int_0^T \int_\Omega (R + \eta) \mathbf{u} \cdot \partial_t \tilde{\mathbf{q}} - \int_0^T \int_\Omega \partial_t \eta \mathbf{u} \cdot \tilde{\mathbf{q}}$$

$$- \int_\Omega (R + \eta_0) \mathbf{u}_0 \cdot \tilde{\mathbf{q}}(0),$$

where we recall that $\tilde{\mathbf{q}} = \mathbf{q} \circ \mathcal{A}^\eta$.

Thus, we have shown that the limiting functions $\mathbf{u}$ and η satisfy the weak form of the coupled, continuous problem 2.3.12, for all test functions $(\tilde{\mathbf{q}}, \psi)$, which are dense in the test space $\widetilde{Q}^\eta$. Thus, he following theorem holds:

Theorem 2.3 (Main Theorem)   *Let ϱ_f, ϱ_s, μ, h, C_i, $i = 1, 2, 3$. Suppose that the initial data $v_0 \in L^2(0, L)$, $\mathbf{u}_0 \in L^2(\Omega_{\eta_0})$, and $\eta_0 \in H_0^2(0, L)$ are such that $(R + \eta_0(z)) > 0$, $z \in [0, L]$. Furthermore, let P_{in}, $P_{out} \in L^2_{loc}(0, \infty)$.*

Then there exist a $T > 0$ and a weak solution $(\mathbf{u}, \eta)$ on $(0, T)$ in the sense of Definition 2.2 (or equivalently Definition 2.1), which satisfies the following energy estimate:

$$E(t) + \int_0^t D(\tau) d\tau \leq E_0 + C(\|P_{in}\|^2_{L^2(0,t)} + \|P_{out}\|^2_{L^2(0,t)}), \quad t \in [0, T],$$

$$(2.7.4)$$

where C depends only on the coefficients in the problem, E_0 is the kinetic energy of the initial data, and $E(t)$ and $D(t)$ are given by

$$E(t) = \frac{\rho_f}{2} \|\mathbf{u}\|^2_{L^2(\Omega^\eta(t))} + \frac{\rho_s h}{2} \|\partial_t \eta\|^2_{L^2(\Gamma))}$$

$$+ \frac{1}{2}\left(C_0 \|\eta\|^2_{L^2(\Gamma)} + C_1 \|\partial_z \eta\|^2_{L^2(\Gamma)} + C_2 \|\partial_z^2 \eta\|^2_{L^2(\Gamma)}\right),$$

$$D(t) = \mu \|\mathbf{D}(\mathbf{u})\|^2_{L^2(\Omega^\eta(t)))} + D_0 \|\partial_t \eta\|^2_{L^2(\Gamma)} + D_1 \|\partial_t \partial_z \eta\|^2_{L^2(\Gamma)} + D_2 \|\partial_t \partial_z^2 \eta\|^2_{L^2(\Gamma)}.$$

Furthermore, one of the following is true: either

1. $T = \infty$, or

2. $\lim_{t \to T} \min_{z \in [0, L]} (R + \eta(z)) = 0$.

Proof It only remains to prove the last assertion, which states that our result is either global in time, or, in the case the walls of the cylinder touch each other, our existence result holds until the time of degeneracy. This can be proved by using similar argument as in [21, p. 397–398]. For the sake of completeness we present the arguments here.

Let $(0, T_1)$, $T_1 > 0$, be the interval on which we have constructed our solution $(\mathbf{u}, \eta)$, and let $m_1 = \min_{(0,T_1) \times (0,L)} (R + \eta)$. From Lemma 2.5 we know that $m_1 >$

0. Furthermore, since $\eta \in W^{1,\infty}(0, T; L^2(0, L)) \cap L^\infty(0, T; H_0^2(0, L))$ and $\mathbf{u} \in L^\infty(0, T; L^2(\Omega))$, we can take T_1 such that $\eta(T_1) \in H_0^2(0, L)$, $\partial_t \eta(T_1) \in L^2(0, L)$ and $\mathbf{u}(T_1) \in L^2(\Omega)$. We can now use the first part of Theorem 2.3 to prolong the solution $(\mathbf{u}, \eta)$ to the interval $(0, T_2)$, $T_2 > T_1$. By iteration, we can continue the construction of our solution to the interval $(0, T_k)$, $k \in \mathbb{N}$, where $(T_k)_{k \in \mathbb{N}}$ is an increasing sequence. We set $m_k = \min_{(0, T_k) \times (0, L)} (R + \eta) > 0$.

Since $m_k > 0$ we can continue the construction further. Without loss of generality we could choose a $T_{k+1} > T_k$ so that $m_{k+1} \geq \frac{m_k}{2}$. From (2.7.4) and from the embedding

$$L^\infty(0, T; H_0^2(0, L)) \cap W^{1,\infty}(0, T; L^2(0, L)) \hookrightarrow C^{0,1-\alpha}([0, T]; H^{2\alpha}(0, L)),$$

by taking $\alpha = 1/2$, we have that the displacement η is Hölder continuous in time, namely,

$$\|\eta\|_{C^{0,1/2}(0, T_{k+1}; C[0, L])} \leq C(T_{k+1}).$$

Therefore, the following estimate holds:

$$R + \eta(T_{k+1}, z) \geq R + \eta(T_k, z) - C(T_{k+1})(T_{k+1} - T_k)^{1/2} \geq m_k - C(T_{k+1})(T_{k+1} - T_k)^{1/2}.$$

For a T_{k+1} chosen so that $m_{k+1} \geq \frac{m_k}{2}$ this estimate implies

$$T_{k+1} - T_k \geq \frac{m_k^2}{4C(T_{k+1})^2}, \quad k \in \mathbb{N}. \tag{2.7.5}$$

Now, let us take $T^* = \sup_{k \in \mathbb{N}} T_k$ and set $m^* = \min_{(0, T^*) \times (0, L)} (R + \eta)$. Obviously, $m_k \geq m^*$, $k \in \mathbb{N}$. There are two possibilities. Either $m^* = 0$, or $m^* > 0$. If $m^* = 0$, this means that $\lim_{t \to T} \min_{z \in [0, L]} (R + \eta(z)) = 0$, and the second statement in the theorem is proved. If $m^* > 0$, we need to show that $T^* = \infty$. To do that, notice that (2.7.4) gives the form of the constant $C(T)$ which is a non-decreasing function of T. Therefore, we have $C(T_k) \leq C(T^*)$, $\forall k \in \mathbb{N}$. Using this observation and that fact that $m_k \geq m^*$, $k \in \mathbb{N}$, estimate (2.7.5) implies

$$T_{k+1} - T_k \geq \frac{(m^*)^2}{2C(T^*)^2}, \quad \forall k \in \mathbb{N}.$$

Since this holds for all $k \in \mathbb{N}$, we have that $T^* = \infty$.

2.7.3 A Few Final Remarks About the Problem

Strong Convergence of $\partial_t \eta_N$ Because we are working with divergence free velocities, we need strong convergence of the $\partial_t \eta_N$'s to be able to get strong convergence of the $\mathbf{u}_N$'s. There are two reasons for that. The first is the fact that we use the strong convergence of the η_N's in our compactness argument for $\mathbf{u}$ (this may possibly be avoided for compressible flows). The second reason is that although we have strong convergence of the $\mathbf{u}_N$'s to $\mathbf{u}$ in $L^2(0, T, L^2(\Omega))$, the trace of the limiting function is not defined in general for such functions. However, in the case of *divergence-free* vector fields, the *normal trace* of $\mathbf{u}$ *is defined*, and needs to be equal to $\partial_t \eta$ in the limit. Therefore, we need strong convergence of the $\partial_t \eta_N$'s to $\partial_t \eta$ to be able to satisfy the kinematic coupling condition in the limit, stating that the normal trace of $\mathbf{u}$ on the moving boundary equals $\partial_t \eta$.

Radial Displacement Assuming only radial (normal) displacement to be different from zero is an assumption which is reasonable for problems for which the structure displacement is primarily a consequence of the pressure loading (since pressure is the main component in the normal stress).

2D Versus 3D (Lipschitz Interface Regularity) One difference between the 2D result presented here and 3D problems is the fact that in 3D, the limiting η is not Lispchitz in the sense that the Lipschitz property does not follow from energy estimates. Our limiting interface is Hölder continuous in the sense that it belongs to the space $H^2(\Gamma)$. When $\Gamma \subset \mathbb{R}$, the interface is Lipschitz, but not in 2D. In the case of 2D structures interacting with a 3D fluid, $H^2 + \epsilon$ regularity would provide a Lipschitz interface, but we do not have that extra regularity just by looking at the energy of the problem. Muha et al. [61], showed recently that one can, in fact, show that the limiting function will be Lipschitz, but not directly from energy estimates.

One consequence of the non-Lipschitz property of the limiting interface η is that $\mathbf{u} \circ \mathscr{A}^\eta$ is not an H^1 function. This gives rise to various difficulties in the construction of extensions to the maximal domain, including the construction of $\mathbf{q}_1$ test function in the limiting problem.

Acknowledgments This work was supported in part by the National Science Foundation under grants DMS-1853340 and DMS-1613757.

The results presented in these lectures were obtained in collaboration with Prof. Boris Muha (University of Zagreb). Special thanks are extended to Prof. Roland Glowinski of the University of Houston for his contributions and support, to Prof. Martina Bukač (University of Notre Dame) for her collaboration on numerical method development, and to PhD student Marija Galić (University of Zagreb) for her thoughtful contributions. Additional thanks are extended to the PhD students at UC Berkeley, Jeffrey Kuan and Mitchell Taylor, for their careful reading of the manuscript and for their useful comments and suggestions.

References

1. R.A. Adams, J.J. F. Fournier, in *Sobolev Spaces*. Pure and Applied Mathematics (Amsterdam), vol. 140, 2nd edn. (Elsevier/Academic Press, Amsterdam, 2003)
2. M. Astorino, J.-F. Gerbeau, O. Pantz, K.-F. Traoré, Fluid-structure interaction and multi-body contact: application to aortic valves. Comput. Methods Appl. Mech. Eng. **198**(45), 3603–3612 (2009)
3. F.P.T. Baaijens, A fictitious domain/mortar element method for fluid-structure interaction. Int. J. Numer. Meth. Fl. **35**(7), 743–761 (2001)
4. S. Badia, F. Nobile, C. Vergara, Fluid-structure partitioned procedures based on Robin transmission conditions. J. Comput. Phys. **227**, 7027–7051 (2008)
5. V. Barbu, Z. Grujić, I. Lasiecka, A. Tuffaha, Existence of the energy-level weak solutions for a nonlinear fluid-structure interaction model, in *Fluids and Waves*. Contemporary Mathematics, vol. 440 (American Mathematical Society, Providence, 2007), pp. 55–82
6. V. Barbu, Z. Grujić, I. Lasiecka, A. Tuffaha, Smoothness of weak solutions to a nonlinear fluid-structure interaction model. Indiana Univ. Math. J. **57**(3), 1173–1207 (2008)
7. M. Boulakia, Existence of weak solutions for the motion of an elastic structure in an incompressible viscous fluid. C. R. Math. Acad. Sci. Paris **336**(12), 985–990 (2003)
8. M. Bukac, S. Canic, R. Glowinski, B. Muha, A. Quaini, A modular, operator-splitting scheme for fluid-structure interaction problems with thick structures. Int. J. Numer. Methods Fluids **74**(8), 577–604 (2014)
9. M. Bukac, S. Canic, R. Glowinski, J. Tambaca, A. Quaini, Fluid-structure interaction in blood flow capturing non-zero longitudinal structure displacement. J. Comput. Phys. **235**, 515–541 (2013)
10. M. Bukac, S. Canic, B. Muha, R. Glowinski, An operator splitting approach to the solution of fluid structure interaction in hemodynamics, in *Splitting Methods in Communication, Imaging, Science, and Engineering*, ed. by R. Glowinski, S. Osher, Y. Yin. Springer Series in Scientific Computation (Springer, Cham, 2016)
11. M. Bukač, B. Muha, Stability and convergence analysis of the extensions of the kinematically coupled scheme for the fluid-structure interaction. SIAM J. Numer. Anal. **54**(5), 3032–3061 (2016)
12. M. Bukač, I. Yotov, R. Zakerzadeh, P. Zunino, Partitioning strategies for the interaction of a fluid with a poroelastic material based on a Nitsche's coupling approach. Comput. Methods Appl. Mech. Eng. **292**, 138–170 (2015)
13. M. Bukač, I. Yotov, P. Zunino, An operator splitting approach for the interaction between a fluid and a multilayered poroelastic structure. Numer. Methods Partial Differ. Equ. **31**(4), 1054–1100 (2015)
14. M. Bukač, S. Čanić, B. Muha, A nonlinear fluid-structure interaction problem in compliant arteries treated with vascular stents. Appl. Math. Optim. **73**(3), 433–473 (2016)
15. S. Canic, Recent progress on moving boundary problems. Am. Math. Soc. Current Event Bull. (Denver, CO, 2020)
16. S. Canic, New mathematics for next generation stent design. SIAM News 52(3) (2019).
17. S. Canic, B. Muha, M. Bukac, Fluid-structure interaction in hemodynamics: modeling, analysis, and numerical simulation, in *Fluid-Structure Interaction and Biomedical Applications*. Advances in Mathematical Fluid Mechanics (Birkhauser, Basel, 2014)
18. S. Canic, M. Galic, B. Muha, Analysis of a nonlinear moving boundary 3D fluid–stent–shell interaction problem. Z. Angew. Math. Phys. **70**, 44 (2019)
19. S. Canic, M. Galic, B. Muha, J. Tambaca, Analysis of a linear 3D fluid–stent–shell interaction problem. Z. Angew. Math. Phys. **70**(2), 1–38 (2019)
20. P. Causin, J.F. Gerbeau, F. Nobile, Added-mass effect in the design of partitioned algorithms for fluid-structure problems. Comput. Methods Appl. Mech. Eng. **194**(42–44), 4506–4527 (2005)

21. A. Chambolle, B. Desjardins, M.J. Esteban, C. Grandmont, Existence of weak solutions for the unsteady interaction of a viscous fluid with an elastic plate. J. Math. Fluid Mech. **7**(3), 368–404 (2005)

22. C.H. Arthur Cheng, D. Coutand, S. Shkoller, Navier–Stokes equations interacting with a nonlinear elastic biofluid shell. SIAM J. Math. Anal. **39**(3), 742–800 (2007)

23. C.H. Arthur Cheng, S. Shkoller, The interaction of the 3D Navier–Stokes equations with a moving nonlinear Koiter elastic shell. SIAM J. Math. Anal. **42**(3), 1094–1155 (2010)

24. C. Conca, H. Jorge San Martín, M. Tucsnak, Motion of a rigid body in a viscous fluid. C. R. Acad. Sci. I. Math. **328**(6), 473–478 (1999)

25. J. Austin Cottrell, T.J.R. Jughes, Y Bazilevs, *Isogeometric Analysis: Toward Integration of CAD and FEA* (Wiley, London, 2009)

26. D. Coutand, S. Shkoller, Motion of an elastic solid inside an incompressible viscous fluid. Arch. Ration. Mech. Anal. **176**(1), 25–102 (2005)

27. D. Coutand, S. Shkoller, The interaction between quasilinear elastodynamics and the Navier–Stokes equations. Arch. Ration. Mech. Anal. **179**(3), 303–352 (2006)

28. P. Cumsille, T. Takahashi, Wellposedness for the system modelling the motion of a rigid body of arbitrary form in an incompressible viscous fluid. Czechoslovak Math. J. **58**(4), 961–992 (2008)

29. H. da Veiga, On the existence of strong solutions to a coupled fluid-structure evolution problem. J. Math. Fluid Mech. **6**(1), 21–52 (2004)

30. B. Desjardins, M.J. Esteban, Existence of weak solutions for the motion of rigid bodies in a viscous fluid. Arch. Ration. Mech. Anal. **146**(1), 59–71 (1999)

31. B. Desjardins, M.J. Esteban, C. Grandmont, P. Le Tallec, Weak solutions for a fluid-elastic structure interaction model. Rev. Mat. Comput. **14**(2), 523–538 (2001)

32. J. Donéa, A Taylor–Galerkin method for convective transport problems, in *Numerical Methods in Laminar and Turbulent Flow (Seattle, 1983)* (Pineridge, Swansea, 1983), pp. 941–950

33. J. Donea, A. Huerta, J.P. Ponthot, A. Rodriguez-Ferran, in *Arbitrary Lagrangian–Eulerian Method*. Encyclopedia of Computational Mathematics (Wiley, London, 2004)

34. M. Dreher, A Jüngel, Compact families of piecewise constant functions in $L^p(0, T; B)$. Nonlinear Anal. Theory Methods Appl. **75**(6), 3072–3077 (2012)

35. Q. Du, M.D. Gunzburger, L.S. Hou, J. Lee, Analysis of a linear fluid-structure interaction problem. Discrete Contin. Dyn. Syst. **9**(3), 633–650 (2003)

36. C. Farhat, P. Geuzaine, C. Grandmont, The discrete geometric conservation law and the nonlinear stability of ale schemes for the solution of flow problems on moving grids. J. Comput. Phys. **174**, 669–694 (2001)

37. L.J. Fauci, R. Dillon, Biofluidmechanics of reproduction. Ann. Rev. Fluid Mech. **38**, 371–394 (2006)

38. E. Feireisl, On the motion of rigid bodies in a viscous compressible fluid. Arch. Ration. Mech. Anal. **167**(4), 281–308 (2003)

39. C.A. Figueroa, I.E. Vignon-Clementel, K.E. Jansen, T.J.R. Hughes, C.A. Taylor, A coupled momentum method for modeling blood flow in three-dimensional deformable arteries. Comput. Methods Appl. Mech. Eng. **195**(41–43), 5685–5706 (2006)

40. G.P. Galdi, in *An Introduction to the Mathematical Theory of the Navier–Stokes Equations I: Linearized Steady Problems*. Springer Tracts in Natural Philosophy, vol. 38 (Springer, New York, 1994)

41. G.P. Galdi, Mathematical problems in classical and non-Newtonian fluid mechanics, in *Hemodynamical Flows*. Oberwolfach Seminars, vol. 37 (Birkhäuser, Basel, 2008), pp. 121–273

42. L. Gerardo-Giorda, F. Nobile, C. Vergara, Analysis and optimization of robin-robin partitioned procedures in fluid-structure interaction problems. SIAM J. Numer. Anal. **48**(6), 2091–2116 (2010)

43. R. Glowinski, Finite element methods for incompressible viscous flow, in *Handbook of Numerical Analysis*. Handbook of Numerical Analysis, vol. IX (North-Holland, Amsterdam, 2003), pp. 3–1176

44. C. Grandmont, M. Hillairet, Existence of global strong solutions to a beam-fluid interaction system. Arch. Ration. Mech. Anal. **220**(3). 1283–1333 (2016)
45. C. Grandmont, M. Lukáčová-Medvid'ová, Š. Nečasová, Mathematical and numerical analysis of some FSI problems, in *Fluid-Structure Interaction and Biomedical Applications*, ed. by T. Bodnár, G. P. Galdi, Š. Nečasová. Advances in Mathematical Fluid Mechanics (Birkhäuser, Basel, 2014)
46. C. Grandmont, Existence of weak solutions for the unsteady interaction of a viscous fluid with an elastic plate. SIAM J. Math. Anal. **40**(2), 716–737 (2008)
47. B.E. Griffith, R.D. Hornung, D.M. McQueen, C.S. Peskin, An adaptive, formally second order accurate version of the immersed boundary method. J. Comput. Phys. **223**(1), 10–49 (2007)
48. T. Hughes, W. Liu, T. Zimmermann, Lagrangian–Eulerian finite element formulation for incompressible viscous flows. Comput. Methods Appl. Mech. Eng. **29**(3), 329–349 (1981)
49. M. Ignatova, I.I. Kukavica, I. Lasiecka, A. Tuffaha, On well-posedness and small data global existence for an interface damped free boundary fluid-structure model. Nonlinearity **27**(3), 467 (2014)
50. M. Ignatova, I. Kukavica, I. Lasiecka, A. Tuffaha, On well-posedness for a free boundary fluid-structure model. J. Math. Phys. **53**(11), 115624, 13 (2012)
51. M. Krafczyk, M. Cerrolaza, M. Schulz, E. Rank, Analysis of 3D transient blood flow passing through an artificial aortic valve by Lattice–Boltzmann methods. J. Biomech. **31**(5), 453–462 (1998)
52. I. Kukavica, A. Tuffaha, Well-posedness for the compressible Navier–Stokes–Lamé system with a free interface. Nonlinearity **25**(11), 3111 (2012)
53. I. Kukavica, A. Tuffaha, M. Ziane, Strong solutions for a fluid structure interaction system. Adv. Differ. Equ. **15**(3–4), 231–254 (2010)
54. I. Kukavica, A. Tuffaha, Solutions to a fluid-structure interaction free boundary problem. Discrete Continuous Dyn. Syst. A **32**(4), 1355–1389 (2012)
55. O.A. Ladyzhenskaya, Initial-boundary problem for Navier–Stokes equations in domains with time-varying boundaries, in *Boundary Value Problems of Mathematical Physics and Related Aspects of Function Theory* (Springer, Berlin, 1970), pp. 35–46
56. P. Le Tallec, J. Mouro, Fluid structure interaction with large structural displacements. Comput. Methods Appl. Mech. Eng. **190**(24–25), 3039–3067 (2001)
57. D. Lengeler, M. Ružička, Weak solutions for an incompressible Newtonian fluid interacting with a Koiter type shell. Arch. Rational Mech. Anal. **211**, 205–255 (2014)
58. J. Lequeurre, Existence of strong solutions to a fluid-structure system. SIAM J. Math. Anal. **43**(1), 389–410 (2011)
59. M. Lukáčová-Medvid'ová, G. Rusnáková, A. Hundertmark-Zaušková, Kinematic splitting algorithm for fluid-structure interaction in hemodynamics. Comput. Methods Appl. Mech. Eng. **265**, 83–106 (2013)
60. S.E. Mikhailov, Traces, extensions and co-normal derivatives for elliptic systems on Lipschitz domains. J. Math. Anal. Appl. **378**, 324–342 (2011)
61. B. Muha, S. Schwarzacher, Existence and regularity for weak solutions for a fluid interacting with a non-linear shell in 3D (2019, submitted)
62. B. Muha, S. Čanić, Existence of a weak solution to a nonlinear fluid-structure interaction problem modeling the flow of an incompressible, viscous fluid in a cylinder with deformable walls. Arch. Ration. Mech. Anal. **207**(3), 919–968 (2013)
63. B. Muha, S. Čanić, A nonlinear, 3D fluid-structure interaction problem driven by the time-dependent dynamic pressure data: a constructive existence proof. Commun. Inf. Syst. **13**(3), 357–397 (2013)
64. B. Muha, S. Čanić, Existence of a solution to a fluid-multi-layered-structure interaction problem. J. Differ. Equ. **256**(2), 658–706 (2014)
65. B. Muha, S. Čanić, Fluid-structure interaction between an incompressible, viscous 3D fluid and an elastic shell with nonlinear Koiter membrane energy. Interfaces Free Bound. **17**(4), 465–495 (2015)

66. B. Muha, S. Čanić, Existence of a weak solution to a fluid-elastic structure interaction problem with the Navier slip boundary condition. J. Differ. Equ. **260**(12), 8550–8589 (2016)
67. B. Muha, S. Čanić, A generalization of the Aubin–Lions–Simon compactness lemma for problems on moving domains. J. Differ. Equ. **266**(12), 8370–8418 (2019)
68. P. Nägele, *Monotone Operator Theory for Unsteady Problems on Non-Cylindrical Domains*, Dissertation. Albert-Ludwigs-Universität Freiburg, Freiburg, 2015
69. C.S. Peskin, The immersed boundary method. Acta Nume. **11**, 479–517 (2002)
70. C.S. Peskin, D.M. McQueen, A three-dimensional computational method for blood flow in the heart I. Immersed elastic fibers in a viscous incompressible fluid. J. Comput. Phys. **81**(2), 372–405 (1989)
71. A. Quarteroni, M. Tuveri, A. Veneziani, Computational vascular fluid dynamics: problems, models and methods. Survey article. Comput. Visual. Sci. **2**, 163–197 (2000)
72. J-P. Raymond, M. Vanninathan, A fluid-structure model coupling the Navier–Stokes equations and the Lamé system. J. Math. Pures Appl. (9) **102**(3), 546–596 (2014)
73. T. Richter, T Wick, Finite elements for fluid-structure interaction in ALE and fully Eulerian coordinates. Comput. Methods Appl. Mech. Eng. **199**, 2633–2642 (2010)
74. J. San Martín, V. Starovoitov, M. Tucsnak, Global weak solutions for the two-dimensional motion of several rigid bodies in an incompressible viscous fluid. Arch. Rational Mech. Anal. **161**(2), 113–147 (2002)
75. R Temam, *Navier–Stokes Equations. Theory and Numerical Analysis*. Studies in Mathematics and its Applications, vol. 2 (North-Holland, Amsterdam, 1977)
76. I. Velčić, Nonlinear weakly curved rod by γ-convergence. J. Elasticity **108**(2), 125–150 (2012)
77. T Wick, Fluid-structure interactions using different mesh motion techniques. Comput. Struct. **89**(1314), 1456–1467 (2011)

Chapter 3
Regularity and Inviscid Limits in Hydrodynamic Models

Peter Constantin

Abstract We discuss the vanishing viscosity limit and low regularity bounds, uniform in viscosity, for vorticity in Yudovich class in 2D. We also show that multiscale steady solutions of Navier–Stokes equations with power law energy spectrum, including K41, can be constructed in any domain in 3D

3.1 Introduction

The three-dimensional incompressible Navier–Stokes equations are the basic equations of mathematical fluid mechanics. The equations

$$\partial_t u + u \cdot \nabla u + \nabla p - \nu \Delta u = f, \qquad (3.1.1)$$

with the incompressibility constraint

$$\nabla \cdot u = 0, \qquad (3.1.2)$$

describe the motion of a fluid of uniform density (taken above to be identically 1), with velocity $u = u(x, t) \in \mathbb{R}^d$ with $x \in \mathbb{R}^d$, $t \geq 0$, in $d = 2$ or $d = 3$ dimensions. The scalar unknown $p = p(x, t)$ represents the hydrodynamic pressure, arising in response to the constraint of incompressibility (3.1.2). The positive number ν represents the kinematic viscosity, and f are body forces.

The Euler equations,

$$\partial_t u + u \cdot \nabla u + \nabla p = f, \qquad (3.1.3)$$

P. Constantin (✉)
Department of Mathematics, Princeton University, Princeton, NJ, USA
e-mail: const@math.princeton.edu

© The Editor(s) (if applicable) and The Author(s), under exclusive licence to Springer Nature Switzerland AG 2020
L. C. Berselli, M. Růžička (eds.), *Progress in Mathematical Fluid Dynamics*, Lecture Notes in Mathematics 2272, https://doi.org/10.1007/978-3-030-54899-5_3

together with the incompressibility condition (3.1.2) are obtained by formally setting $\nu = 0$ in the Navier–Stokes equations. The pressure enforces the incompressibility condition, and if the forces are divergence-free, the pressure must satisfy

$$- \Delta p = \nabla \cdot (u \cdot \nabla u) . \tag{3.1.4}$$

The subject of these lectures is motivated by questions arising in turbulence, one of the greatest challenges in physics. A law of turbulence states that the average rate of dissipation of kinetic energy per unit mass does not vanish in the limit of infinite Reynolds numbers.

$$- \lim_{Re \to \infty} \langle \frac{dE}{dt} \rangle = \epsilon > 0$$

This law is experimentally well verified. Another important law of turbulence theory is the K41 spectrum, or Kolmogorov–Obukhov spectrum,

$$E(k) = C \epsilon^{\frac{2}{3}} k^{-\frac{5}{3}},$$

which states that the energy per wave number k has a universal power law behavior for a range of scales, called the inertial range. This range extends from low wave numbers, where the energy injection typically occurs, to a viscosity dependent cutoff wave number, which converges to infinity in the limit of zero viscosity. This again is very well verified experimentally. The physical literature on the subject is vast. A lucid presentation is given in [1].

The mathematical description of these two laws requires a more precise formulation. The laws are not in any way mathematical statements, and formulations can be given so that they invalid. The more challenging task is to understand why they are observed in nature, and how are they related to the fundamental underlying equations. In these lectures we present negative results, results in which the vanishing viscosity limit is conservative, and results in which non-turbulent Navier–Stokes stationary solutions exhibit power law scaling behavior.

3.2   Inviscid Limit

If we consider the issue of the limit of energy dissipation, we certainly can find cases in which the limit vanishes. These are cases in which the solutions of the Navier–Stokes equations converge to solutions of Euler equations, and the latter are smooth enough to conserve energy. This situation occurs, as it is very well known, if we are considering spatially periodic solutions and solutions of the Euler equations which belong to $H^s(\mathbb{T}^d)$, $s > \frac{d}{2} + 1$ [2, 3].

The difference between solutions vanishes in the inviscid limit, in strong norms, at a rate proportional to the difference between coefficients, that is, linearly with viscosity. This rate changes if we consider less smooth solutions of Euler equations, even in 2D. This was first investigated in [4] and [5] for vortex patches, a class of weak solutions of Euler equations in 2D. We describe below recent results [6] extending the earlier work.

3.2.1 Yudovich Class

We discuss here the connection between Yudovich solutions of the Euler equations [27]

$$\partial_t \omega + u \cdot \nabla \omega = g, \tag{3.2.1}$$

with bounded forcing $g \in L^\infty(0, T; L^\infty(\mathbb{T}^2))$, and initial data

$$\omega(0) = \omega_0 \in L^\infty(\mathbb{T}^2), \tag{3.2.2}$$

and the vanishing viscosity limit $(\lim_{\nu \to 0})$ of solutions of the Navier–Stokes equations,

$$\partial_t \omega^\nu + u^\nu \cdot \nabla \omega^\nu = \nu \Delta \omega^\nu + g, \tag{3.2.3}$$

with initial data

$$\omega^\nu(0) = \omega_0^\nu \in L^\infty(\mathbb{T}^2), \tag{3.2.4}$$

and the same forcing g. We consider uniformly bounded initial data

$$\sup_{\nu > 0} \|\omega_0^\nu\|_{L^\infty(\mathbb{T}^2)} \leq \Omega_{0,\infty} < \infty. \tag{3.2.5}$$

The solutions of (3.2.1), (3.2.2), (3.2.3), and (3.2.4) are uniformly bounded in $L^\infty(\mathbb{T}^2)$:

$$\sup_{\nu \geq 0} \sup_{0 \leq t \leq T} \|\omega^\nu(t)\|_{L^\infty(\mathbb{T}^2)} \leq \Omega_\infty = \Omega_{0,\infty} + \int_0^T \|g(t)\|_{L^\infty(\mathbb{T}^2)} dt. \tag{3.2.6}$$

This bound is valid in $\mathbb{T}^2$ or $\mathbb{R}^2$ but is not available if boundaries are present or in 3D. The bound will be used repeatedly below.

The vorticity distribution function $\pi_{\omega^\nu(t)}(dy)$ is defined by

$$\int f(y)\pi_{\omega^\nu(t)}(dy) = \int f(\omega^\nu(t, x))dx, \tag{3.2.7}$$

for all continuous functions (observables) f. If $\omega_0^\nu \to \omega_0$ we the distributions convergence

$$\pi_{\omega^\nu(t)}(dy) \xrightarrow{\nu \to 0} \pi_{\omega(t)}(dy) = \pi_{\omega_0}(dy), \qquad (3.2.8)$$

where the time invariance of the vorticity distribution function for the Euler equations follows from Lagrangian transport $\omega(t) = \omega_0 \circ X_t^{-1}$ and volume preservation of the homeomorphism $A_t = X_t^{-1}$. The statement (3.2.8) is a consequence of the strong convergence of the vorticity in $L^\infty(0, T; L^p(\mathbb{T}^2))$ for all $p \in [1, \infty)$ and for any $T > 0$. This fact was proved in [6], extending previous work for vortex patch solutions with smooth boundary [4], and removing additional assumptions on the Euler path [5]. This result has implications for equilibrium theories [28] of decaying two-dimensional turbulence [7, 8, 29] The result of [6] is:

Theorem 3.1 *Let ω be the unique Yudovich weak solution of the Euler equations with initial data $\omega_0 \in L^\infty(\mathbb{T}^2)$ and forcing $g \in L^\infty(0, T; L^\infty(\mathbb{T}^2))$. Let ω^ν be the solution of the Navier–Stokes equation with the same forcing and initial data $\omega_0^\nu \to \omega_0$ strongly in $L^2(\mathbb{T}^2)$. Then, for any $T > 0$ and $p \in [1, \infty)$, the inviscid limit $\omega^\nu \to \omega$ holds strongly in $L^\infty(0, T; L^p(\mathbb{T}^2))$:*

$$\lim_{\nu \to 0} \sup_{0 \le t \le T} \|\omega^\nu(t) - \omega(t)\|_{L^p(\mathbb{T}^2)} = 0. \qquad (3.2.9)$$

Consequently, the distributions converge,

$$\lim_{\nu \to 0} \pi_{\omega^\nu(t)}(dy) = \pi_{\omega_0}(dy), \qquad (3.2.10)$$

for all $t \in [0, T]$.

Remark 3.1 The result is sharp, in several ways. First, there can be no infinite time result as the Euler solution is conservative and the Navier–Stokes solution is dissipative. Secondly, there can be no rate without additional regularity assumptions on ω_0, as is the case for the heat equation. Thirdly, there can be no strong convergence in L^∞ because ω_0 may not be continuous while ω^ν is smooth for any $t > 0$. And, finally there can be no strong convergence for $p > 1$ in domains with boundaries, if the boundary condition of the Navier–Stokes solutions is no slip, and the Euler solution has non-vanishing tangential velocity at the boundary, in other words, if there are boundary layers [9].

The method of proof of Theorem 3.1 yielded also the continuity of the Yudovich solution map $\omega(t) = S(t)(\omega_0)$ in the L^p topology when restricted to fixed balls in L^∞.

Theorem 3.2 *For any* $\omega_0, \omega_0^n \in L^\infty(\mathbb{T}^2)$ *such that* ω_0^n *is uniformly bounded in* $L^\infty(\mathbb{T}^2)$ *and* $\omega_0^n \to \omega_0$ *as* $n \to \infty$ *strongly in* $L^2(\mathbb{T}^2)$ *we have*

$$\lim_{n \to \infty} \|S(t)(\omega_0^n) - S(t)(\omega_0)\|_{L^p(\mathbb{T}^2)} = 0 \tag{3.2.11}$$

for each time $t > 0$.

If additional smoothness is assumed on the data then some degree of fractional smoothness in L^p can be propagated uniformly in viscosity [6]:

Theorem 3.3 *Suppose* $\omega_0 \in (L^\infty \cap B_{p,\infty}^s)(\mathbb{T}^2)$ *for some* $s > 0$ *and some* $p \geq 1$. *Then the solutions of the Navier–Stokes equations satisfy* $\omega^\nu(t) \in (L^\infty \cap B_{p,\infty}^{s(t)})(\mathbb{T}^2)$ *uniformly in* ν, *where*

$$s(t) = s \exp(-Ct\|\omega_0\|_{L^\infty(\mathbb{T}^2)})$$

for some universal constant $C > 0$.

The proof of Theorem 3.3 relied on the fact that the velocity is log-Lipschitz uniformly in ν and showed that the exponential estimate with loss of [10] holds uniformly in viscosity. The proof given in [6] used the stochastic Lagrangian representation formula of [11]

$$dX_t(x) = u^\nu(X_t(x), t)dt + \sqrt{2\nu}\, dW_t, \qquad X_0(x) = x, \tag{3.2.12}$$

yielding the representation formula

$$\omega^\nu(t) = \mathbb{E}\left[\omega_0 \circ A_t\right] \tag{3.2.13}$$

where back-to-labels map is defined as $A_t = X_t^{-1}$. The noisy Lagrangian picture allowed for adaptation of ideas of [10, 12] to the viscous case. Uniform Sobolev regularity could be established by similar arguments; if $\omega_0 \in (L^\infty \cap W^{s,p})(\mathbb{T}^2)$ then $\omega^\nu(t) \in (L^\infty \cap W^{s(t),p})(\mathbb{T}^2)$ with uniformly bounded norms.

The uniform regularity of Theorem 3.3 was used to deduce

Corollary 3.1 *Let* $\omega_0 \in (L^\infty \cap B_{2,\infty}^s)(\mathbb{T}^2)$ *with* $s > 0$ *and let* ω *and* ω^ν *solve respectively (3.2.1) and (3.2.3), with the same initial data* $\omega_0^\nu = \omega_0$. *Then the* L^p *convergence of vorticity, for any* $p \in [1, \infty)$ *and any finite time* $T > 0$, *occurs at the rate*

$$\sup_{t \in [0,T]} \|\omega^\nu(t) - \omega(t)\|_{L^p(\mathbb{T}^2)} \lesssim (\nu T)^{\frac{s \exp(-2CT\|\omega_0\|_\infty)}{p(1+s\exp(-CT\|\omega_0\|_\infty))}-}, \tag{3.2.14}$$

with the universal constant $C > 0$ *in Proposition 3.3.*

Corollary 3.1 applies in particular to the to inviscid limits of vortex patches with non-smooth boundary. Indeed, Lemma 3.2 of [5] shows that if $\omega_0 = \chi_\Omega$ is the characteristic function of a bounded domain whose boundary has box-counting (fractal) dimension D not larger than the dimension of space $d = 2$, i.e. $d_F(\partial\Omega) := D < 2$, then $\omega_0 \in B^{(2-D)/p}_{p,\infty}(\mathbb{T}^2)$. Proposition 3.3 then shows that some degree of fractional Besov regularity of the solution $\omega^\nu(t)$ is retained uniformly in viscosity for any finite time $T < \infty$ and Corollary 3.1 provides a rate depending only D, T and p at which the vanishing viscosity limit holds, removing therefore the need for the additional assumptions on the solution imposed in [5].

The proof of Theorem 3.1, adapted from [6], is given below. It is based on a number of properties of Yudovich class solutions, in particular the exponential integrability of gradients and the fact that linear transport by Yudovich solutions has a short time uniformly controlled loss of regularity: it maps bounded sets in $W^{1,p}$, $p > 2$ to bounded sets in H^1, uniformly in viscosity.

We give further a proof of a uniform propagation of regularity result, Theorem 3.4, a version of Theorem 3.3 which does not use the stochastic representation.

We start the proof of Theorem 3.1 with the exponential integrability of gradients of velocities obtained via the Biot–Savart law in dimension two.

Lemma 3.1 *Let $\omega \in L^\infty(\mathbb{T}^2)$ and let u be obtained from ω by the Biot–Savart law*

$$u = K[\omega] = \nabla^\perp(\Delta)^{-1}\omega. \tag{3.2.15}$$

There exists a non-dimensional constant $\gamma > 0$ and a constant C_K with units of area such that

$$\int_{\mathbb{T}^2} \exp\{\beta|\nabla u(x)|\}\,dx \le C_K \tag{3.2.16}$$

holds for any $\beta > 0$ such that

$$\beta\|\omega\|_{L^\infty(\mathbb{T}^2)} \le \gamma. \tag{3.2.17}$$

Proof The bound (3.2.16) holds due to the fact that Calderon–Zygmund operators map L^∞ to BMO [13], $\omega \in L^\infty \mapsto \nabla u = \nabla K[u] \in BMO$, and from the John–Nirenberg inequality [14] for BMO functions. We provide below a direct and elementary argument (modulo a fact about norms of singular integral operators), for the sake of completeness.

We recall that there exists a constant C_* so that for all $p \ge 2$,

$$\|\nabla K[v]\|_{L^p(\mathbb{T}^2)} = \|\nabla \otimes \nabla(-\Delta)^{-1}v\|_{L^p(\mathbb{T}^2)} \le C_* p\|v\|_{L^p(\mathbb{T}^2)}. \tag{3.2.18}$$

(See [13]). The dependence of (3.2.18) on p is the important point. Thus,

$$\int_{\mathbb{T}^2} e^{\beta |\nabla u|} \mathrm{d}x = \sum_{p=0}^{\infty} \beta^p \frac{\|\nabla u\|_{L^p(\mathbb{T}^2)}^p}{p!}$$

$$\leq \sum_{p=0}^{\infty} \frac{\left(C_* \beta \|\omega\|_{L^p(\mathbb{T}^2)}\right)^p p^p}{p!} \leq |\mathbb{T}^2| \sum_{p=0}^{\infty} \frac{\left(C_* \beta \|\omega\|_{L^\infty(\mathbb{T}^2)}\right)^p p^p}{p!}.$$

This is a convergent series provided $C_* \beta \|\omega\|_{L^\infty(\mathbb{T}^2)} < 1/e$. Indeed, this can be seen using Stirling's bound $n! \geq \sqrt{2\pi} n^{n+1/2} e^{-n}$ which yields

$$\sum_{p=0}^{\infty} \frac{c^p p^p}{p!} \leq 1 + \sum_{p=1}^{\infty} \frac{p^{-1/2}}{\sqrt{2\pi}} (ce)^p \leq \frac{1}{1-ce}, \quad \text{provided} \quad c \in [0, 1/e)$$

where $c := C_* \beta \|\omega\|_{L^\infty(\mathbb{T}^2)}$. In (3.2.16) we may take thus

$$\gamma = (2C_* e)^{-1}, \quad C_K = 2\left|\mathbb{T}^2\right|. \tag{3.2.19}$$

The constant γ depends on the Biot–Savart kernel and is non-dimensional, the constant C_K then is proportional to the area of the domain.

The next result establishes strong convergence of the velocity in $L^2(0, T; L^2(\mathbb{T}^2))$. If $g = 0$ and $u_0^\nu = u_0$, this is a consequence of Theorem 1.4 of [15]. Below is a generalization of [15] which applies in our setting and is proved by a different argument.

Lemma 3.2 *Let* $\omega_0 \in L^\infty(\mathbb{T}^2)$. *There exist constants* U, Ω_2 *and* K *(see below (3.2.23), (3.2.24), and (3.2.39)) depending on norms of the initial data and of the forcing such that the difference* $v = u^\nu - u$ *of velocities of solutions (3.2.1) and (3.2.3) obeys*

$$\|v(t)\|_{L^2}^2 \leq 3U^2 K^{\frac{5(t-t_0)\Omega_\infty}{\gamma}} \left(\frac{\|v(t_0)\|_{L^2(\mathbb{T}^2)}^2}{U^2} + \gamma \frac{\Omega_2^2}{U^2 \Omega_\infty} \nu \right)^{1 - \frac{5(t-t_0)\Omega_\infty}{\gamma}} \tag{3.2.20}$$

for all $0 \leq t_0 \leq t$. *By iterating the above, we obtain*

$$\|v(t)\|_{L^2}^2 \leq 20U^2 K^{1 - e^{-10t\Omega_\infty/\gamma}} \left(\frac{\|v(0)\|_{L^2(\mathbb{T}^2)}^2}{U^2} + \gamma \frac{\Omega_2^2}{U^2 \Omega_\infty} \nu \right)^{e^{-\frac{10t\Omega_\infty}{\gamma}}} \tag{3.2.21}$$

provided that $\|v(0)\|_{L^2(\mathbb{T}^2)}^2 + \gamma \nu \Omega_2^2 / \Omega_\infty \leq 9KU^2$.

Remark 3.2 (Continuity of Solution Map) At zero viscosity, Lemma 3.2 establishes Hölder continuity of the Yudovich (velocity) solution map. Specifically, denoting $S(t)(u_0)$ the velocity with initial data u_0 and $\nu = 0$, a consequence of Lemma 3.2

is that $\|S(t)v(u_0) - S(t)(u_0')\|_{L^2(\mathbb{T}^2)} \leq C\|u_0 - u_0'\|_{L^2(\mathbb{T}^2)}^{\alpha(t)}$ where $\alpha(t) := e^{-ct}$ and $c, C > 0$ are appropriate constants. This fact is used to prove Theorem 3.2. It is worth further remarking that the condition on the data $\|v(0)\|_{L^2(\mathbb{T}^2)}^2 \leq 9KU^2$ required for the above estimate to hold is $O(1)$ (data need not be taken very close).

Proof The proof of Lemma 3.2 proceeds in two steps.

Step 1: Short Time Bound The proof of the lemma starts from the equation obeyed by the difference v,

$$\partial_t v + u^\nu \cdot \nabla v + v \cdot \nabla u + \nabla p = \nu \Delta v + \nu \Delta u$$

leading to the inequality

$$\frac{d}{dt}\|v\|_{L^2}^2 + \nu\|\nabla v\|_{L^2}^2 \leq \nu\|\nabla u\|_{L^2}^2 + 2\int |\nabla u||v|^2 \mathrm{d}x \tag{3.2.22}$$

which is a straightforward consequence of the equation, using just integration by parts. We use the bound Ω_∞ (3.2.6) for the vorticity of the Euler solution. We also use a bound for the L^2 norms

$$\sup_{0 \leq t \leq T} \left(\|u^\nu(t)\|_{L^2(\mathbb{T}^2)} + \|u(t)\|_{L^2(\mathbb{T}^2)}\right) \leq U, \tag{3.2.23}$$

which is easily obtained from energy balance. We use also bounds for L^p norms of vorticity,

$$\Omega_p = \sup_{0 \leq t \leq T} \|\omega(t)\|_{L^p(\mathbb{T}^2)}. \tag{3.2.24}$$

We split the integral

$$\int |\nabla u||v|^2 \mathrm{d}x = \int_B |\nabla u||v|^2 \mathrm{d}x + \int_{\mathbb{T}^2 \setminus B} |\nabla u||v|^2 \mathrm{d}x$$

where

$$B = \{x \mid |v(x, t)| \geq MU\}$$

with M to be determined below. Although B depends in general on time, it has small measure if M is large,

$$|B| \leq M^{-2}.$$

The constant M has dimensions of inverse length. We bound

$$2 \int_B |\nabla u||v|^2 \mathrm{d}x \leq 2\|\nabla u\|_{L^2} \|v\|_{L^4}^2 \leq 2|B|^{\frac{1}{4}} \|\nabla u\|_{L^4} \|v(t)\|_{L^4}^2 \qquad (3.2.25)$$

where we used $\int_B |\nabla u|^2 \mathrm{d}x \leq |B|^{\frac{1}{2}} \|\nabla u\|_{L^4}^2$. We now use the fact that we are in Yudovich class and Ladyzhenskaya inequality to deduce

$$\|v(t)\|_{L^4}^2 \leq C\|v(t)\|_{L^2}[\|\omega_0\|_{L^2} + \|g\|_{L^1(0,T;L^2)}] \leq CU\Omega_2$$

and we use also

$$\|\nabla u\|_{L^4} \leq [C\|\omega_0\|_{L^4} + \|g\|_{L^1(0,T;L^4)}] = \Omega_4$$

to bound (3.2.25) by

$$2 \int_B |\nabla u||v|^2 \mathrm{d}x \leq CU\Omega_2\Omega_4 M^{-\frac{1}{2}}, \qquad (3.2.26)$$

We non-dimensionalize by dividing by U^2 and we multiply by $\beta = \gamma/\Omega_\infty$. The quantity

$$y(t) = \frac{\|v(t)\|_{L^2(\mathbb{T}^2)}^2}{U^2} \qquad (3.2.27)$$

obeys the inequality

$$\beta \frac{\mathrm{d}y}{\mathrm{d}t} \leq \beta v \frac{\Omega_2^2}{U^2} + C\beta\Omega_4 \frac{\Omega_2}{U} M^{-\frac{1}{2}} + 2 \int_{\mathbb{T}^2 \setminus B} \beta |\nabla u| \frac{|v|^2}{U^2} \mathrm{d}x. \qquad (3.2.28)$$

We write the term

$$2 \int_{\mathbb{T}^2 \setminus B} \beta |\nabla u||v|^2 U^{-2} \mathrm{d}x = 2 \int_{\mathbb{T}^2 \setminus B} (\beta |\nabla u| + \log \epsilon + \log \frac{1}{\epsilon})|v|^2 U^{-2} \mathrm{d}x \qquad (3.2.29)$$

with ϵ (with units of inverse area) to be determined below. We use the inequality (3.2.56) and Lemma 3.1 with

$$a = \beta |\nabla u| + \log \epsilon, \qquad b = \frac{|v|^2}{U^2}$$

to deduce

$$2 \int_{\mathbb{T}^2 \setminus B} \beta |\nabla u||v|^2 U^{-2} \mathrm{d}x \leq 2\epsilon C_K + 2 \log \frac{M^2}{\epsilon} y(t). \qquad (3.2.30)$$

Inserting (3.2.30) in (3.2.28) we obtain

$$\beta \frac{dy}{dt} \le F + \log\left(\frac{M^2}{\epsilon}\right) y(t) \tag{3.2.31}$$

with

$$F = \beta v \frac{\Omega_2^2}{U^2} + C\beta \Omega_4 \frac{\Omega_2}{U} M^{-\frac{1}{2}} + 2\epsilon C_K. \tag{3.2.32}$$

Note that F and $\frac{M^2}{\epsilon}$ are non-dimensional. From (3.2.31) we obtain immediately

$$y(t) \le \left(\frac{M^2}{\epsilon}\right)^{\frac{t-t_0}{\beta}} y(t_0) + \frac{F}{\log\left(\frac{M^2}{\epsilon}\right)} \left(\left(\frac{M^2}{\epsilon}\right)^{\frac{t-t_0}{\beta}} - 1\right). \tag{3.2.33}$$

We choose M such that

$$C\beta\Omega_4 \frac{\Omega_2}{U} M^{-\frac{1}{2}} = \beta v \frac{\Omega_2^2}{U^2} + y(t_0) \tag{3.2.34}$$

and we choose ϵ such that

$$2\epsilon C_K = \beta v \frac{\Omega_2^2}{U^2} + y(t_0). \tag{3.2.35}$$

These choices imply

$$F = 3\beta v \frac{\Omega_2^2}{U^2} + 2y(t_0). \tag{3.2.36}$$

Then we see that

$$\Gamma = \frac{M^2}{\epsilon} = 2C_K \left(C\beta\Omega_4 \frac{\Omega_2}{U}\right)^4 \times \left(\beta v \frac{\Omega_2^2}{U^2} + y(t_0)\right)^{-5}. \tag{3.2.37}$$

Taking without loss of generality $\log \Gamma \ge 1$, we have from (3.2.33)

$$\begin{aligned}
y(t) &\le 3\left(y(t_0) + \beta v \frac{\Omega_2^2}{U^2}\right) \Gamma^{\frac{t-t_0}{\beta}} \\
&\le 3\left(y(t_0) + \beta v \frac{\Omega_2^2}{U^2}\right)^{1 - \frac{5(t-t_0)}{\beta}} \times \left(2C_K \left(C\beta\Omega_4 \frac{\Omega_2}{U}\right)^4\right)^{\frac{5(t-t_0)}{\beta}}.
\end{aligned} \tag{3.2.38}$$

Recalling that $\beta = \gamma/\Omega_\infty$ and denoting the non-dimensional constant

$$K = 2C_K \left(C\beta\Omega_4 \frac{\Omega_2}{U} \right)^4 \tag{3.2.39}$$

we established

$$\frac{\|v(t)\|^2}{U^2} \leq 3K^{\frac{5(t-t_0)\Omega_\infty}{\gamma}} \left(\frac{\|v(t_0)\|^2_{L^2(\mathbb{T}^2)}}{U^2} + \beta v \frac{\Omega_2^2}{U^2} \right)^{1 - \frac{5(t-t_0)\Omega_\infty}{\gamma}}. \tag{3.2.40}$$

Thus, we established (3.2.20).

Step 2: Long Time Bound With (3.2.20) established, we now prove (3.2.21). Let $c = 5\Omega_\infty/\gamma$, $\Delta t = 1/2c$ and $t_i = t_{i-1} + \Delta t$ and $a_i = \|v(t_i)\|^2_{L^2}/U^2$ for $i \in \mathbb{N}$. Then (3.2.20) states

$$a_i \leq C_1 (a_{i-1} + C_2 v)^{1/2}, \qquad i = 1, 2, \ldots \tag{3.2.41}$$

with $C_1 = 3K^{\frac{5\Omega_\infty}{2c\gamma}} = 3K^{\frac{1}{2}}$ and $C_2 = \beta \frac{\Omega_2^2}{U^2}$. We set

$$\delta_n = \frac{a_i + C_2 v}{C_1^2} \tag{3.2.42}$$

and observe that (3.2.41) is

$$\delta_n \leq \sqrt{\delta_{n-1}} + \tilde{v} \tag{3.2.43}$$

where

$$\tilde{v} = \frac{C_2 v}{C_1^2} \tag{3.2.44}$$

is a non-dimensional inverse Reynolds number. It follows then by induction that

$$\delta_n \leq (\delta_0)^{2^{-n}} + \sum_{i=0}^{n-1} (\tilde{v})^{2^{-i}}. \tag{3.2.45}$$

Indeed, the induction step follows from

$$\delta_{n+1} \leq \sqrt{\delta_n} + \tilde{v} \tag{3.2.46}$$

and the subadditivity of $\lambda \mapsto \sqrt{\lambda}$. If

$$\tilde{v} \leq \frac{1}{\sqrt{5} - 1} \tag{3.2.47}$$

then the iteration (3.2.43) starting from $0 < \delta_0 < r$ where r is the positive root of the equation $x^2 - x - \tilde{v} = 0$, remains in the interval $(0, r)$, and for any n, δ_n obeys (3.2.45). We observe that

$$\sum_{i=0}^{n-1}(\tilde{v})^{2^{-i}} = (\tilde{v})^{2^{-n+1}}\left(1 + \cdots + (\tilde{v})^{2^{n-1}}\right) \leq \frac{1}{1-\tilde{v}}(\tilde{v})^{2^{-n+1}} \tag{3.2.48}$$

and therefore (3.2.21) follows from (3.2.45). We note that the iteration defined with equality in (3.2.43) converges as $n \to \infty$ to r. Fixing any $t > 0$ and letting $n = \lceil t/\Delta t \rceil = \lceil 2ct \rceil = \lceil 10t\Omega_\infty/\gamma \rceil$ establishes the bound.

The next useful result concerns scalars transported and amplified by a velocity with bounded curl in two dimensions.

Lemma 3.3 *Let* $u := u(x, t)$ *be divergence free and* $\omega := \nabla^\perp \cdot u \in L^\infty(0, T; L^\infty(\mathbb{T}^2))$ *with*

$$\sup_{0 \leq t \leq T} \|\omega(t)\|_{L^\infty(\mathbb{T}^2)} \leq \Omega_\infty. \tag{3.2.49}$$

Consider a nonnegative scalar field $\theta := \theta(x, t)$ *satisfying the differential inequality*

$$\partial_t \theta + u \cdot \nabla\theta - \nu\Delta\theta \leq |\nabla u|\theta + f, \tag{3.2.50}$$

with initial data $\theta|_{t=0} = \theta_0 \in L^\infty(\mathbb{T}^2)$, *and forcing* $f \in L^\infty(0, T; L^\infty(\mathbb{T}^2))$. *Let* $\gamma > 0$ *be the constant from Lemma 3.1. Then, for any* $p > 1$ *and the time* $T(p) = \frac{\gamma(p-1)}{2p\Omega_\infty}$ *it holds that*

$$\sup_{t \in [0, T(p)]} \|\theta(t)\|_{L^2(\mathbb{T}^2)} \leq C_1\|\theta_0\|_{L^{2p}(\mathbb{T}^2)}^p + C_2 \tag{3.2.51}$$

for some constants C_1, C_2 *depending only on* p, Ω_∞ *and* $\|f\|_{L^\infty(0,T;L^\infty(\mathbb{T}^2))}$.

Proof Let $p := p(t)$ with $p(0) = p_0$ and time dependence of $p(t)$ to be specified below. Consider

$$\frac{1}{2}\frac{d}{dt}\int_{\mathbb{T}^2}|\theta|^{2p(t)}dx = p'(t)\int_{\mathbb{T}^2}\ln|\theta||\theta|^{2p(t)}dx + p(t)\int_{\mathbb{T}^2}|\theta|^{2p(t)-2}\theta\partial_t\theta dx$$

$$\leq p'(t)\int_{\mathbb{T}^2}\ln|\theta||\theta|^{2p(t)}dx - p(t)\int_{\mathbb{T}^2}|\theta|^{2p(t)-2}\theta u \cdot \nabla\theta dx$$

$$+ \nu p(t)\int_{\mathbb{T}^2}|\theta|^{2p(t)-2}\theta \Delta\theta dx + p(t)\int_{\mathbb{T}^2}|\theta|^{2p(t)-2}|\nabla u|\theta^2 dx$$

$$+ p(t)\int_{\mathbb{T}^2}|\theta|^{2p(t)-2}\theta f dx. \tag{3.2.52}$$

We now use the following facts

$$\int_{\mathbb{T}^2} |\theta|^{2p-2}\theta f \, dx \leq C\|f\|_{L^\infty(0,T;L^\infty(\mathbb{T}^2))} \|\theta\|_{2p}^{2p-1}, \tag{3.2.53}$$

$$p \int_{\mathbb{T}^2} |\theta|^{2p-2}\theta u \cdot \nabla\theta \, dx = \frac{1}{2}\int_{\mathbb{T}^2} u \cdot \nabla(|\theta|^{2p})dx = 0, \tag{3.2.54}$$

$$\nu \int_{\mathbb{T}^2} |\theta|^{2p-2}\theta \Delta\theta \, dx = -\nu(2p-1)\int_{\mathbb{T}^2} |\theta|^{2p-2}|\nabla\theta|^2 dx \leq 0. \tag{3.2.55}$$

In the second equality we used the fact that the velocity is divergence free. Altogether we find thus

$$\frac{1}{2}\frac{d}{dt}\|\theta(t)\|_{2p(t)}^{2p(t)}dx$$
$$\leq p'(t)\int_{\mathbb{T}^2} \ln|\theta||\theta|^{2p(t)}dx + p(t)\int_{\mathbb{T}^2}|\theta|^{2p(t)}|\nabla u|dx + p(t)\|f\|_{L^\infty}\|\theta\|_{2p}^{2p-1}.$$

We now use the following elementary inequality: for $a, b > 0$,

$$ab \leq e^a + b \ln b - b. \tag{3.2.56}$$

In fact, we use only that $ab \leq e^a + b \ln b$. The inequality (3.2.56) is proved via calculus and follows because the Legendre transform of the convex function $b \ln b - b + 1$ is $e^a - 1$. Setting $a = \beta|\nabla u|$ and $b = \frac{1}{\beta}|\theta|^{2p}$, applying (3.2.56) and Lemma 3.1 we obtain

$$\frac{1}{2}\frac{d}{dt}\|\theta(t)\|_{2p(t)}^{2p(t)} \leq p'(t)\int_{\mathbb{T}^2}\ln|\theta||\theta|^{2p}dx + \frac{p(t)}{\beta}\int_{\mathbb{T}^2}\ln(\beta^{-1}|\theta|^{2p})|\theta|^{2p}dx$$
$$+ p(t)\int_{\mathbb{T}^2}e^{\beta|\nabla u|}dx + Cp(t)\|f\|_{L^\infty}\|\theta\|_{2p}^{2p-1}$$
$$\leq \left(p'(t) + \frac{2p(t)^2}{\beta}\right)\int_{\mathbb{T}^2}\ln|\theta||\theta|^{2p}dx + \frac{p(t)}{\beta}\ln(\beta^{-1})\|\theta(t)\|_{2p}^{2p}$$
$$+ p(t)C_K + Cp(t)\|f\|_{L^\infty}\|\theta\|_{2p}^{2p-1}, \tag{3.2.57}$$

where C_K is the constant from Lemma 3.1 and $\beta = \frac{\gamma}{\Omega_\infty}$ depends on the bound for $\|\omega(t)\|_{L^\infty}$. We now choose p to evolve according to

$$p'(t) = -2\beta^{-1}p(t)^2, \quad p(0) = p_0 \implies p(t) = \frac{\beta p_0}{\beta + 2p_0 t}. \tag{3.2.58}$$

Note that $p(t)$ is a positive monotonically decreasing function of t. Let the time t_* defined by $t_* = T(p_0) := \beta(p_0 - 1)/2p_0$ be such that $p(t_*) = 1$. Then $p(t) \in [1, p_0]$ for all $t \in [0, t_*]$. Note also from (3.2.58) that

$$\int_0^t p(s)ds = \log\left(\frac{p_0}{p(t)}\right)^{2\beta} = \log\left(1 + \frac{2p_0 t}{\beta}\right)^{\frac{2}{\beta}}.$$

Defining $m(t) = \frac{1}{2}\|\theta(t)\|_{2p(t)}^{2p(t)}$ and using (3.2.58) we have the differential inequality

$$m'(t) \le p(t)(C_1 m(t) + C_2) \implies C_1 m(t) + C_2 = (C_1 m_0 + C_2)\left(1 + \frac{2p_0 t}{\beta}\right)^{\frac{2C_1}{\beta}}$$

$$(3.2.59)$$

with C_1 and C_2 depending on $\|f\|_{L^\infty(0,T;L^\infty(\mathbb{T}^2))}$, p_0, C_K and β. Thus

$$m(t) \le m_0\left(1 + \frac{2p_0 t}{\beta}\right)^{\frac{2C_1}{\beta}} + \frac{C_2}{C_1}\left[\left(1 + \frac{2p_0 t}{\beta}\right)^{\frac{2C_1}{\beta}} - 1\right]$$

Note that $p_0/p(t) = 1 + 2p_0\beta^{-1}t$ is increasing on $[0, t_*]$ from 1 to $p_0/p(t_*) = p_0$. Consequently

$$\|\theta(t)\|_{2p(t)} \le C_1\|\theta_0\|_{2p_0}^{p_0} + C_2 \qquad (3.2.60)$$

where the constants C_1 and C_2 have been redefined but the dependence on parameters is the same. As $p(t) \in [1, p_0]$ for all $t \in [0, t_*]$ we have that $\|\theta(t)\|_2 \le \|\theta(t)\|_{2p(t)}$ and we obtain

$$\sup_{t \in [0,t_*]} \|\theta(t)\|_2 \le C_1\|\theta_0\|_{2p_0}^{p_0} + C_2, \qquad (3.2.61)$$

which completes the proof.

We prove now a short time inviscid limit result, in which the time of convergence importantly depends only on L^∞ initial vorticity bounds.

Proposition 3.1 *Let ω and ω^ν solve (3.2.1) and (3.2.3) respectively, with initial data (3.2.2) and (3.2.4). Assume that the Navier–Stokes initial data converge uniformly in $L^2(\mathbb{T}^2)$*

$$\lim_{\nu \to 0} \|\omega_0^\nu - \omega_0\|_{L^2(\mathbb{T}^2)} = 0. \qquad (3.2.62)$$

Assume also that there exists a constant Ω_∞ such that the initial data are uniformly bounded in $L^\infty(\mathbb{T}^2)$:

$$\sup_{\nu>0} \|\omega_0^\nu\|_{L^\infty(\mathbb{T}^2)} \leq \Omega_\infty. \tag{3.2.63}$$

Then there exists a constant C_* such that the vanishing viscosity limit holds

$$\lim_{\nu\to 0} \sup_{t\in[0,T_*]} \|\omega^\nu(t) - \omega(t)\|_{L^2(\mathbb{T}^2)} = 0 \tag{3.2.64}$$

on the time interval $[0, T_*]$ where

$$T_* = (C_*\Omega_\infty)^{-1}. \tag{3.2.65}$$

Proof For the proof we introduce functions ω_ℓ and ω_ℓ^ν which are the unique solutions of the following *linear* problems. We fix $\ell > 0$ and let

$$\partial_t\omega_\ell + u \cdot \nabla\omega_\ell = \varphi_\ell * g, \qquad \omega_\ell(0) = \varphi_\ell * \omega_0, \tag{3.2.66}$$

$$\partial_t\omega_\ell^\nu + u^\nu \cdot \nabla\omega_\ell^\nu = \nu\Delta\omega_\ell^\nu + \varphi_\ell * g, \qquad \omega_\ell^\nu(0) = \varphi_\ell * \omega_0^\nu, \tag{3.2.67}$$

where φ_ℓ is a standard mollifier at scale ℓ and where u and u^ν are respectively the unique solutions of Euler and Navier–Stokes equations. Note that the solutions to the linear problems (3.2.66) and (3.2.67) exist globally and are unique because the Yudovich velocity field u is log-Lipshitz. We observe that we have

$$\|\omega^\nu(t) - \omega(t)\|_{L^2(\mathbb{T}^2)}$$
$$\leq \|\omega(t) - \omega_\ell(t)\|_{L^2(\mathbb{T}^2)} + \|\omega^\nu(t) - \omega_\ell^\nu(t)\|_{L^2(\mathbb{T}^2)} + \|\omega_\ell^\nu(t) - \omega_\ell(t)\|_{L^2(\mathbb{T}^2)}. \tag{3.2.68}$$

Because the equations for $\omega_\ell, \omega_\ell^\nu$ and, respectively ω, ω^ν share the same incompressible velocities, we find

$$\|\omega(t) - \omega_\ell(t)\|_{L^2(\mathbb{T}^2)} \leq \|\omega_0 - \varphi_\ell * \omega_0\|_{L^2(\mathbb{T}^2)} + \int_0^t \|g(s) - \varphi_\ell * g(s)\|_{L^2(\mathbb{T}^2)}ds, \tag{3.2.69}$$

$$\|\omega^\nu(t) - \omega_\ell^\nu(t)\|_{L^2(\mathbb{T}^2)} \leq \|\omega_0^\nu - \varphi_\ell * \omega_0^\nu\|_{L^2(\mathbb{T}^2)} + \int_0^t \|g(s) - \varphi_\ell * g(s)\|_{L^2(\mathbb{T}^2)}ds. \tag{3.2.70}$$

As mollification can be removed strongly in L^p, the two terms in the right hand sides converge to zero as $\ell, \nu \to 0$, in any order.

It remains to show that

$$\lim_{\nu \to 0} \sup_{t \in [0,T_*]} \|\omega_\ell^\nu(t) - \omega_\ell(t)\|_{L^2(\mathbb{T}^2)} \to 0 \qquad (3.2.71)$$

for fixed ℓ. We show now that the two-dimensional linearized Euler and Navier–Stokes equations have uniformly bounded vorticity gradients for short time. This is done in the following Lemma.

Lemma 3.4 *Fix $\ell > 0$ and let ω_ℓ and ω_ℓ^ν solve (3.2.66) and (3.2.67) respectively. Then there exists a constant C_* and a constant $C_\ell < \infty$ depending only on ℓ, the forcing norm $\|g\|_{L^\infty(0,T;L^\infty(\mathbb{T}^2))}$, and the uniform bound on solutions given in (3.2.6) such that for $T_* \leq (C_*\Omega_\infty)^{-1}$, we have that*

$$\sup_{t \in [0,T_*]} \left(\|\omega_\ell(t)\|_{H^1} + \|\omega_\ell^\nu(t)\|_{H^1} \right) \leq C_\ell. \qquad (3.2.72)$$

For the proof of this lemma we provide a viscosity independent bound for $\|\omega_\ell^\nu(t)\|_{H^1}$. The proof for $\|\omega_\ell(t)\|_{H^1}$ is the same, setting $\nu = 0$. We show that $|\nabla \omega_\ell^\nu|$ obeys (3.2.50). Differentiating (3.2.67), we find

$$(\partial_t + u^\nu \cdot \nabla)\nabla \omega_\ell^\nu + \nabla u^\nu \cdot \nabla \omega_\ell^\nu = \nu\Delta(\nabla \omega_\ell^\nu) + \nabla(\varphi_\ell * g). \qquad (3.2.73)$$

A standard computation shows that $|\nabla \omega_\ell^\nu|$ satisfies

$$(\partial_t + u^\nu \cdot \nabla - \nu\Delta)|\nabla \omega_\ell^\nu| \leq |\nabla u||\nabla \omega_\ell^\nu| + |\nabla(\varphi_\ell * g)| \qquad (3.2.74)$$

which is a particular case of the scalar inequality (3.2.50) with $\theta = |\nabla \omega_\ell^\nu|$, initial data $\theta_0 = |\nabla(\varphi_\ell * \omega_0^\nu)| \in L^\infty(\mathbb{T}^2)$ and forcing $f = |\nabla(\varphi_\ell * g)| \in L^\infty(0, T; L^\infty(\mathbb{T}^2))$, as claimed. Applying Lemma 3.3, we find that for any $p > 1$ (e.g. $p = 2$) we have

$$\sup_{t \in [0,T_*]} \|\omega_\ell^\nu(t)\|_{H^1} = C_1 \frac{1}{\ell^p}\left(\int_{\mathbb{T}^2} |\omega_0^\nu * (\nabla\varphi)_\ell|^{2p}dx\right)^{1/2} + C_2 \qquad (3.2.75)$$
$$\leq C_\ell \|\omega_0^\nu\|_{L^\infty(\mathbb{T}^2)}^p \leq C_\ell \Omega_\infty^p.$$

The constant C_ℓ diverges with the mollification scale ℓ, through the prefactor ℓ^{-p} and through the dependence on $\|\nabla(\varphi_\ell * g)\|_{L^\infty} \lesssim \ell^{-1}\|g\|_{L^\infty}$. The important point however is that (3.2.75) holds uniformly in viscosity, completing the proof of Lemma 3.4. Using it, the difference enstrophy obeys

$$\frac{d}{dt}\|\omega_\ell^\nu - \omega_\ell\|_{L^2(\mathbb{T}^2)}^2 = -\int_{\mathbb{T}^2}(u^\nu - u) \cdot \nabla \omega_\ell^\nu(\omega_\ell^\nu - \omega_\ell)dx - \nu\int_{\mathbb{T}^2}|\nabla \omega_\ell^\nu|^2 dx$$
$$+\nu\int_{\mathbb{T}^2}\nabla \omega_\ell^\nu \cdot \nabla\omega_\ell dx \leq 4\Omega\|u^\nu - u\|_{L^2(\mathbb{T}^2)}\|\omega_\ell^\nu\|_{L^2(\mathbb{T}^2)} + \nu\|\nabla\omega_\ell^\nu\|_{L^2(\mathbb{T}^2)}\|\nabla\omega_\ell\|_{L^2(\mathbb{T}^2)}$$
$$\lesssim C_\ell\|u^\nu - u\|_{L^\infty(0,T;L^2(\mathbb{T}^2))} + \nu C_\ell^2. \qquad (3.2.76)$$

Integrating we find

$$\|\omega_\ell^\nu - \omega_\ell\|_{L^2(\mathbb{T}^2)}^2 \lesssim \|\varphi_\ell * (\omega_0^\nu - \omega_0)\|_{L^2(\mathbb{T}^2)}^2 + C_\ell T \|u^\nu - u\|_{L^\infty(0,T;L^2(\mathbb{T}^2))} + \nu C_\ell^2 T.$$
(3.2.77)

To conclude the proof we must show that, at fixed $\ell > 0$, we have $\lim_{\nu \to 0} \|\omega_\ell^\nu - \omega_\ell\|_{L^2(\mathbb{T}^2)} = 0$. Recall that by our assumption (3.2.62) we have that $\lim_{\nu \to 0} \|\omega_0^\nu - \omega_0\|_{L^2(\mathbb{T}^2)} \to 0$. Due to assumption (3.2.62) we have that $\lim_{\nu \to 0} \|u_0^\nu - u_0\|_{L^2(\mathbb{T}^2)} \to 0$. Lemma 3.2 then allows us to conclude from (3.2.77) that $\lim_{\nu \to 0} \sup_{t \in [0,T_*]} \|\omega_\ell^\nu - \omega_\ell\|_{L^2(\mathbb{T}^2)} \to 0$ at fixed $\ell > 0$ and the proof of Proposition 3.1 is complete.

Proof of Theorem 3.1 It suffices to prove that $\lim_{\nu \to 0} \sup_{t \in [0,T]} \|\omega^\nu(t) - \omega(t)\|_{L^2(\mathbb{T}^2)} = 0$. Indeed, convergence in L^p for any $p \in [2, \infty)$ then follows from interpolation and boundedness in L^∞:

$$\|\omega^\nu(t) - \omega(t)\|_{L^p(\mathbb{T}^2)} \le 2\Omega_\infty^{\frac{p-2}{p}} \|\omega^\nu(t) - \omega(t)\|_{L^2(\mathbb{T}^2)}^{\frac{2}{p}}.$$
(3.2.78)

In order to establish strong $L_t^\infty L_x^2$ convergence for arbitrary finite times T, it is enough to the convergence for a short time which depends only on a uniform L^∞ bound on the initial vorticity. The proof of Theorem 3.1 follows by dividing the time interval $[0, T]$ in subintervals

$$[0, T] = [0, T_*] \cup [T_*, 2T_*] \cup \cdots$$

where T_* is determined from the uniform bound (3.2.6), and applying Proposition 3.1 to each interval, with initial data $\omega(nT_*)$, and respectively $\omega^\nu(nT_*)$. As there is no required rate of convergence for the initial data in Proposition 3.1, Theorem 3.1 follows.

3.2.2 Uniform Regularity

In this section we consider for simplicity the unforced case in $\mathbb{R}^2$. We study propagation of low regularity, uniform in viscosity. Let us consider the Navier–Stokes equation in $\mathbb{R}^2$

$$\partial_t \omega + u \cdot \nabla \omega - \nu \Delta \omega = 0,$$
(3.2.79)

with initial vorticity $\omega_0 \in \mathbb{Y}$ where

$$\mathbb{Y} = L^1(\mathbb{R}^2) \cap L^\infty(\mathbb{R}^2).$$
(3.2.80)

The velocity u is given by the Biot–Savart law, (3.2.15). The main result of this section is the following.

Theorem 3.4 *Let* $1 < p < \infty$. *Let* $\omega_0 \in \Upsilon \cap B_{p,1}^s(\mathbb{R}^2)$. *There exist constants* C_Ω *and* Ω_∞ *depending only on the norm of the initial data in* Υ *such that the solution of the Navier–Stokes equations (3.2.79) with initial data* ω_0 *(3.2.92) satisfies, uniformly in* ν,

$$\|\omega(t)\|_{B_{p,1}^{s(t)}(\mathbb{R}^2)} \leq e^{Ct\Omega_\infty} \|\omega_0\|_{B_{p,1}^s(\mathbb{R}^2)} \tag{3.2.81}$$

with

$$s(t) = s - (5\log 2C_\Omega)t \tag{3.2.82}$$

for $0 \leq t \leq (5\log 2C_\Omega)^{-1} s$.

Remark 3.3 Note that in view of the embeddings

$$B_{p,\infty}^{s'}(\mathbb{R}^n) \subset B_{p,1}^s(\mathbb{R}^n) \subset B_{p,\infty}^s(\mathbb{R}^n)$$

for $0 \leq s < s'$ we can track the regularity of solutions with initial data in $B_{p,\infty}^{s'}(\mathbb{R}^2)$, and hence that of vortex patches with rough boundaries, of positive codimension.

We recall the fact that Biot–Savart velocities of Yudovich class vorticities are log-Lipschitz:

Proposition 3.2 *Let* $u = K[\omega]$ *be given by the Biot–Savart law (3.2.15) and let* $\omega \in \Upsilon$. *There exists a constant* C *such that*

$$|u(x+h) - u(x)| \leq C\Omega_\infty |h| \left[1 + \log\left(1 + \frac{L}{|h|}\right) \right] \tag{3.2.83}$$

holds for $x, h \in \mathbb{R}^2$, *where* $L = \sqrt{\frac{\Omega_1}{\Omega_\infty}}$ *and* Ω_p *are the* $L^p(\mathbb{R}^2)$ *norms of* ω.

Proof We write

$$u(x+h) - u(x) = \int_{\mathbb{R}^2} (k(x-y+h) - k(x-y))\omega(y)dy = \int_{\mathbb{R}^2} (k(z+h) - k(z))\omega(x-z)dz,$$

where

$$k(z) = \frac{1}{2\pi} \frac{z^\perp}{|z|^2}. \tag{3.2.84}$$

We split the integral in two, corresponding to $|z| \leq 2|h|$ and $|z| \geq 2|h|$. We have

$$\left| \int_{|z| \leq 2|h|} |k(z+h)| |\omega(x-z)| dz \right| + \left| \int_{|z| \leq 2|h|} |k(z)| |\omega(x-z)| dz \right| \leq C|h| \|\omega\|_{L^{\infty}(\mathbb{R}^2)}$$

by passing to polar coordinates centered at $-h$ and respectively at 0, and using $|k(x)| \leq \frac{1}{2\pi|x|}$. The second integral we bound by

$$\int_{|z| \geq 2|h|} |k(z+h) - k(z)| |\omega(x-z)| dz$$
$$\leq C|h| \int_0^1 d\lambda \int_{|z+\lambda h| \geq |h|} |z+\lambda h|^{-2} |\omega(x-z)| dz$$

here we used $|\nabla k(x)| \leq C|x|^{-2}$. Now we split the integral again, for $|z + \lambda h| \leq L$ and $|z + \lambda h| \geq L$. In the first integral we use L^{∞} bounds on ω and obtain a logarithmic dependence, $\|\omega\|_{L^{\infty}} \log \frac{L}{|h|}$ and in the second integral we use L^1 bounds on ω and we obtain $L^{-2} \|\omega\|_{L^1}$.

We recall some facts about the Littlewood–Paley decomposition. We start with a smooth, nonincreasing, radial nonnegative function $\phi(r)$ satisfying

$$\begin{cases} \phi(r) = 1, & \text{for } 0 \leq r \leq a, \\ \phi(r) = 0, & \text{for } b \leq r, \\ 0 < a < b. \end{cases}$$

We define

$$\psi_0(r) = \phi\left(\frac{r}{2}\right) - \phi(r),$$

$$(\Delta_{-1}u)(x) = (\phi(D)u)(x) = (2\pi)^{-n} \int_{\mathbb{R}^n} e^{ix \cdot \xi} \phi(|\xi|) \widehat{u}(\xi) d\xi, \tag{3.2.85}$$

$$(\Delta_0 u)(x) = (\psi_0(D)u)(x) = (2\pi)^{-n} \int_{\mathbb{R}^n} e^{ix \cdot \xi} \psi_0(|\xi|) \widehat{u}(\xi) d\xi, \tag{3.2.86}$$

$$\psi_j(r) = \psi_0(2^{-j}r)$$

and

$$(\Delta_j u)(x) = (\psi_j(D)u)(x) = (2\pi)^{-n} \int_{\mathbb{R}^n} e^{ix \cdot \xi} \psi_j(|\xi|) \widehat{u}(\xi) d\xi, \tag{3.2.87}$$

where

$$\widehat{u}(\xi) = \mathcal{F}u(\xi) = \int_{\mathbb{R}^n} e^{-ix \cdot \xi} u(x) dx.$$

We choose $a = \frac{1}{2}, b = \frac{5}{8}$. We set also

$$S_k(u) = \sum_{j=-1}^{k} \Delta_j(u) \qquad (3.2.88)$$

Proposition 3.3 *If $u \in S'(\mathbb{R}^n)$, then*

$$u = \sum_{j=-1}^{\infty} \Delta_j u,$$

$$\mathrm{supp}\mathcal{F}(\Delta_j u) \subset 2^j \left[\frac{1}{2}, \frac{5}{4}\right],$$

for $j \geq 0$, and in particular

$$\Delta_j \Delta_k \neq 0 \Rightarrow |j - k| \leq 1, \quad \text{for } j, k \geq 0.$$

Moreover,

$$(\Delta_j + \Delta_{j+1} + \Delta_{j+2})\Delta_{j+1} = \Delta_{j+1},$$

for $j \geq 0$,

$$\Delta_j (S_{k-2}(u)\Delta_k(v)) \neq 0 \Rightarrow k \in [j - 2, j + 2]$$

for $j \geq 2, k \geq 2$.

Proposition 3.4 (Bernstein Inequalities)

$$\|\Delta_j u\|_{L^q(\mathbb{R}^n)} \leq C 2^{j(\frac{n}{p} - \frac{n}{q})} \|\Delta_j u\|_{L^p(\mathbb{R}^n)}, \quad q \geq p \geq 1,$$

$$\|S_j u\|_{L^q(\mathbb{R}^n)} \leq C 2^{j(\frac{n}{p} - \frac{n}{q})} \|S_j u\|_{L^p(\mathbb{R}^n)}, \quad q \geq p \geq 1,$$

and

$$2^{jm}\|\Delta_j u\|_{L^p(\mathbb{R}^n)} \leq C \sum_{|\alpha|=m} \|\partial^\alpha \Delta_j u\|_{L^p(\mathbb{R}^n)} \leq C 2^{jm}\|\Delta_j u\|_{L^p(\mathbb{R}^n)}$$

We introduce the inhomogeneous Besov space with norm

$$\|u\|_{B^s_{p,q}(\mathbb{R}^n)} = \left\| \left\{ 2^{sj} \|\Delta_j u\|_{L^p(\mathbb{R}^n)} \right\}_j \right\|_{\ell^q(\mathbb{N})}$$

Proposition 3.5 (Littlewood–Paley) *Let* $1 < p < \infty$. *Then* $(\mathbb{I} - \Delta)^{\frac{s}{2}} u \in L^p(\mathbb{R}^n)$ *if and only if* $\Delta_j u \in L^p(\mathbb{R}^n)$ *for all* $j \geq -1$ *and*

$$\|u\|_{W^{s,p}(\mathbb{R}^n)} \sim \left\| \sqrt{\sum_{j \geq -1} 2^{2js} |\Delta_j(u)|^2} \right\|_{L^p(\mathbb{R}^n)}$$

Proposition 3.6 *Embeddings:*

$$B^s_{p,r}(\mathbb{R}^n) \subset B^{s - \left(\frac{n}{p} - \frac{n}{q}\right)}_{q,r}(\mathbb{R}^n), \qquad q \geq p \geq 1,$$

$$B^0_{p,2}(\mathbb{R}^n) \subset L^p(\mathbb{R}^n) \subset B^0_{p,p}(\mathbb{R}^n) \qquad p \geq 2,$$

$$B^0_{p,p}(\mathbb{R}^n) \subset L^p(\mathbb{R}^n) \subset B^0_{p,2}(\mathbb{R}^n) \qquad p \leq 2.$$

Products Consider two functions, $u = \sum_{k \geq -1} \Delta_k u$ and $v = \sum_{l \geq -1} \Delta_l(v)$. Then we have the Bony decomposition

$$\Delta_j(uv) = I_j(u, v) + I_j(v, u) + R_j(u, v) \tag{3.2.89}$$

with

$$I_j(u, v) = \sum_{k \in [j-2, j+2]} \Delta_j(S_{k-2}(u) \Delta_k(v)) \tag{3.2.90}$$

and

$$R_j(u, v) = \sum_{|k-l| \leq 1} \Delta_j(\Delta_k u \Delta_l v). \tag{3.2.91}$$

Proof of Theorem 3.4 We consider the Navier–Stokes vorticity evolution is the $B^s_{p,1}$ space, with $s > 0$ and $1 < p < \infty$. We take initial vorticity

$$\omega_0 \in \mathbb{Y} \cap B^s_{p,1}(\mathbb{R}^2) \tag{3.2.92}$$

and look first at the evolution of $\Delta_j \omega$ in L^p, using the Bony decomposition.

$$\frac{1}{p} \frac{d}{dt} \|\Delta_j \omega\|_{L^p} \leq A_j + B_j + C_j \tag{3.2.93}$$

for $j \geq 5$ where

$$A_j = \left\| \sum_{k \in [j-2,j+2]} \left[S_{k-2}(u), \Delta_j \right] \cdot \nabla \Delta_k \omega \right\|_{L^p(\mathbb{R}^2)}, \tag{3.2.94}$$

$$B_j = \left\| \sum_{k \in [j-2,j+2]} \Delta_j \left(\Delta_k(u), \cdot \nabla S_{k-2}\omega \right) \right\|_{L^p(\mathbb{R}^2)}, \tag{3.2.95}$$

and

$$C_j = \left\| \sum_{|k-l| \leq 1, k \geq j-2} \Delta_j \left(\Delta_l u \cdot \nabla \Delta_k \omega \right) \right\|_{L^p(\mathbb{R}^2)}, \tag{3.2.96}$$

The commutator appears in A_j because of the property $\Delta_j \sum_{k \in [j-2,j+2]} \Delta_k = \Delta_j$ and the fact that $S_{j-2}u$ is divergence-free. We discarded the nonnegative term due to the viscosity. We use the fact that $S_{k-2}(u)$ are uniformly log-Lipschitz:

$$\left| \left[S_{k-2}(u), \Delta_j \right] f(x) \right|$$
$$\leq C\Omega_\infty 2^{2j} \int_{\mathbb{R}^2} |\Psi_0(2^j(x-y))| |x-y| \log \left(1 + \frac{L}{|x-y|} \right) |f(y)| dy$$
$$\leq j 2^{-j} \int_{\mathbb{R}^2} \widetilde{\Psi}(z) |f| (x - 2^{-j}z) dz$$

where Ψ_0 is a Schwartz function, Fourier inverse of ψ_0, $\mathcal{F}\Psi_0 = \psi_0$, and

$$\widetilde{\Psi} = C\Omega_\infty |x| \left(\log \left(1 + \frac{L}{|x|} \right) + \log 2 \right) |\Psi_0(x)| \tag{3.2.97}$$

is rapidly decaying, and hence belongs in $L^1(\mathbb{R}^2)$. Here we used the fact that S_{k-2} commute with translation and are uniformly bounded in all L^p, and hence Ω_∞ and L are bounded independently of k and t.

$$\left\| \left[S_{k-2}(u), \Delta_j \right] f \right\|_{L^p(\mathbb{R}^2)} \leq C_\Omega j 2^{-j} \| f \|_{L^p(\mathbb{R}^2)}$$

and where C_Ω is the L^1 norm of $\widetilde{\Psi}$. It follows that

$$A_j \leq j C_\Omega \sum_{k \in [j-2,j+2]} \| \Delta_k \omega \|_{L^p(\mathbb{R}^2)} \tag{3.2.98}$$

The bound of B_j is more straightforward,

$$B_j \leq C\Omega_\infty \sum_{k \in [j-2,j+2]} \| \Delta_k \omega \|_{L^p(\mathbb{R}^2)} \tag{3.2.99}$$

and uses Bernstein inequalities and the boundedness of ∇K in L^p spaces, where K is the Biot–Savart operator. The remaining term is bounded also using Bernstein inequalities

$$C_j \leq C\Omega_\infty \sum_{l \geq j-3} \|\omega_l\|_{L^p(\mathbb{R}^2)} \tag{3.2.100}$$

We consider now the norm

$$\|\omega(t)\|_{B^{s(t)}_{p,1}} \tag{3.2.101}$$

and arrange the decay of the exponent so that it counter balances the logarithmic growth of the term A_j. In order to do so, we observe that (3.2.98) implies the bound

$$2^{sj}A_j \leq C_\Omega \sum_{k \in [j-2,j+2]} k2^{sk}\|\Delta_k\omega\|_{L^p(\mathbb{R}^2)} \tag{3.2.102}$$

as long as $s \leq 1$, with a slightly larger C_Ω. Similarly, from (3.2.99) and from (3.2.100) we obtain

$$2^{sj}B_j \leq C\Omega_\infty \sum_{k \in [j-2,j+2]} 2^{sk}\|\Delta_k\omega\|_{L^p(\mathbb{R}^2)} \tag{3.2.103}$$

and

$$2^{sj}C_j \leq C\Omega_\infty \sum_{l \geq j-3} 2^{s(j-l)}2^{sl}\|\omega_l\|_{L^p(\mathbb{R}^2)}. \tag{3.2.104}$$

Imposing

$$\frac{ds}{dt} = -5\log 2C_\Omega \tag{3.2.105}$$

where C_Ω is the constant in (3.2.102), we deduce

$$\frac{d}{dt}\|\omega(t)\|_{B^{s(t)}_{p,1}} \leq C\Omega_\infty\|\omega(t)\|_{B^{s(t)}_{p,1}} \tag{3.2.106}$$

This concludes the proof of Theorem 3.4.

3.3   Multiscale Solutions

We describe constructions of solutions of inviscid equations given in [16] which were inspired by work of [17].

For the incompressible 3D Euler equations, if u is Beltrami, i.e. if the curl of the velocity

$$\omega = \nabla \times u \qquad (3.3.1)$$

is parallel to the velocity, and if $u \in L^2(\mathbb{R}^3)$, then u must be identically zero [18, 19]. In fact, Liouville theorems which assert the vanishing of solutions which have constant behavior at infinity are often true for systems of the sort we are discussing. In contrast, vortex rings are examples of solutions of the 3D Euler equations with compactly supported vorticity [20]. However, they have nonzero constant velocities at infinity. Because of the Biot–Savart law

$$u(x,t) = -\frac{1}{4\pi} \int_{\mathbb{R}^3} \frac{x-y}{|x-y|^3} \times \omega(y,t)dy, \qquad (3.3.2)$$

if ω is compactly supported, it is hard to imagine that u can also be compactly supported. In view of these considerations, the following result of Gavrilov [17] was surprising.

Theorem 3.5 *(Gavrilov [17]) There exist nontrivial time independent solutions $u \in \left(C_0^\infty(\mathbb{R}^3)\right)^3$ of the three-dimensional incompressible Euler equations.*

The paper [16] described a construction inspired by the result of Gavrilov but based on Grad–Shafranov equations, classical equations arising in the study of plasmas [21, 22] augmented by a *localizability* condition (see (3.3.17)). This point of view yielded a general method which was applied to several other hydrodynamic equations, revealing a number of universal features. The 3D incompressible Euler equations result which extends Theorem 3.5 is stated in Theorem 3.6. An application providing multiscale steady solutions which are locally smooth, vanish at $\partial\Omega$, but globally belong only to Hölder classes $C^\alpha(\Omega)$ is given in Theorem 3.7. Such solutions can be constructed so that they belong to $L^2(\Omega) \cap C^{\frac{1}{3}}(\Omega)$ but not to any $C^\alpha(\Omega)$ with $\alpha > \frac{1}{3}$, they have vanishing local dissipation $u \cdot \nabla(\frac{|u|^2}{2} + p) = 0$, but have arbitrary large $\||\nabla u||u|^2\|_{L^\infty(\Omega)}$. These solutions conserve energy, as they are stationary in time, and they have the regularity of the dissipative solutions recently constructed in connection with the Onsager conjecture (see review papers [23, 24]). Compactly supported weak solutions which belong to $C^\alpha(\Omega)$ but not to $C^\beta(\Omega)$, $0 < \alpha < \beta \leq 1$ can also be constructed.

3.3.1 Steady Axisymmetric Euler Equations

The stationary 3D axisymmetric Euler equations are solved via the Grad–Shafranov ansatz

$$u = \frac{1}{r}(\partial_z \psi)e_r - \frac{1}{r}(\partial_r \psi)e_z + \frac{1}{r}F(\psi)e_\phi \tag{3.3.3}$$

where $\psi = \psi(r, z)$ is a smooth function of $r > 0$, $z \in \mathbb{R}$, and the swirl F is a smooth function of ψ alone. It is known that smooth compactly supported velocities solving stationary axisymmetric 3D Euler equations must vanish identically if the swirl F vanishes [25]. Above e_r, e_z, e_ϕ are the orthonormal basis of cylindrical coordinates r, z, ϕ with the orientation convention $e_r \times e_\phi = e_z$, $e_r \times e_z = -e_\phi$, $e_\phi \times e_z = e_r$. Note that u is automatically divergence-free,

$$\nabla \cdot u = 0, \tag{3.3.4}$$

and also that, by construction,

$$u \cdot \nabla \psi = 0. \tag{3.3.5}$$

The vorticity $\omega = \nabla \times u$ is given by

$$\omega = -\frac{1}{r}(\partial_z \psi)F'(\psi)e_r + \frac{1}{r}(\partial_r \psi)F'(\psi)e_z + \frac{\Delta^* \psi}{r}e_\phi \tag{3.3.6}$$

where $F' = \frac{dF}{d\psi}$ and the Grad–Shafranov operator Δ^* is

$$\Delta^* \psi = \partial_r^2 \psi - \frac{1}{r}\partial_r \psi + \partial_z^2 \psi. \tag{3.3.7}$$

In view of (3.3.3) and (3.3.6), the vorticity can be written as

$$\omega = -F'(\psi)u + \frac{1}{r}\left(\Delta^* \psi + \frac{1}{2}(F^2)'\right)e_\phi. \tag{3.3.8}$$

As it is very well known, the steady Euler equations

$$u \cdot \nabla u + \nabla p = 0 \tag{3.3.9}$$

can be written as

$$\omega \times u + \nabla\left(\frac{|u|^2}{2} + p\right) = 0, \tag{3.3.10}$$

and therefore the axisymmetric Euler equations are solved if ψ solves the Grad–Shafranov equation [21, 22]

$$- \Delta^* \psi = \partial_\psi \left(\frac{F^2}{2} + r^2 P \right) \tag{3.3.11}$$

where the function $P = P(\psi)$ represents the plasma pressure:

$$\omega \times u = \nabla P. \tag{3.3.12}$$

The analogy with the steady MHD equations $u \leftrightarrow B$, $\omega \leftrightarrow J$ motivates the name. Both the swirl F and the plasma pressure P are arbitrary functions of ψ. The plasma pressure and the hydrodynamic pressure are related via

$$p + \frac{|u|^2}{2} + P = \text{constant}. \tag{3.3.13}$$

The constant should be time independent if we are studying time independent solutions, and it may be taken without loss of generality to be zero.
    If

$$u \cdot \nabla p = 0, \tag{3.3.14}$$

then, together with a solution u, p of (3.3.9, 3.3.4), any function

$$\tilde{u} = \phi(p)u \tag{3.3.15}$$

with ϕ smooth is again a solution of (3.3.9, 3.3.4) with pressure given by

$$\nabla \tilde{p} = \phi^2(p)\nabla p. \tag{3.3.16}$$

This can be used to localize solutions. In his construction Gavrilov obtained solutions u defined in the neighborhood of a circle, obeying the Euler equations near the circle, and having a relationship $|u|^2 = 3p$ between the velocity magnitude and the hydrodynamic pressure.
    This motivates us to consider the overdetermined system formed by the Grad–Shafranov equation for ψ (3.3.11) coupled with the requirement

$$\frac{|u|^2}{2} = A(\psi). \tag{3.3.17}$$

This requirement is the constraint of *localizability* of the Grad–Shafranov equation, and it severely curtails the freedom of choice of functions F and P. This localizability is in fact the essence and the novelty of the method. Without this constraint many solutions (3.3.3) with ψ solving the Grad–Shafranov equation (3.3.11) exist, including explicit ones [26]), but they cannot be localized in space.

The method we are describing consists thus in seeking functions F, P, A of ψ such that the system

$$\begin{cases} -\Delta^*\psi = \partial_\psi\left(\tfrac{1}{2}F^2(\psi) + r^2 P(\psi)\right), \\ |\nabla\psi|^2 + F^2(\psi) = 2r^2 A(\psi), \end{cases} \tag{3.3.18}$$

is solved. Then the function u given in the ansatz (3.3.3), and the pressure

$$p = -P(\psi) - A(\psi) \tag{3.3.19}$$

together satisfy the steady 3D Euler equations (3.3.9, 3.3.4), and are *localizable*, meaning that (3.3.17) is satisfied. It is important to observe that it is enough to find smooth functions F, P, A of ψ and a smooth function ψ in an open set. This open set need not be simply connected, but once u and p are found using this construction, any function $\phi(p)u$ is again a solution of steady Euler equations, and it is sometimes possible to extend this solution to the whole space.

3.3.2 Construction

The construction of solutions of (3.3.18) starts with a hodograph transformation. We seek functions $U(r, \psi)$ and $V(r, \psi)$ defined in an open set in the (r, ψ) plane and a smooth function $\psi(r, z)$ defined in an open set of the (r, z) plane such that the equations

$$\partial_r\psi(r, z) = U(r, \psi(r, z)), \tag{3.3.20}$$

$$\partial_z\psi(r, z) = V(r, \psi(r, z)) \tag{3.3.21}$$

are satisfied. This clearly requires the compatibility

$$V\partial_\psi U = U\partial_\psi V + \partial_r V. \tag{3.3.22}$$

Once the compatibility is satisfied then the solution ψ exists locally (in simply connected components). The system (3.3.18) becomes

$$\begin{cases} \partial_r U + U\partial_\psi U + V\partial_\psi V - \tfrac{1}{r}U = -F\partial_\psi F - r^2\partial_\psi P \\ U^2 + V^2 + F^2 = 2r^2 A. \end{cases} \tag{3.3.23}$$

We traded a system of two equations in two independent variables (r, z) of total degree three, (3.3.18), for a system of three first order equations (3.3.22, 3.3.23) in

two independent variables (r, ψ). We integrate this locally. We start by noticing that
the first equation of (3.3.23) is

$$\partial_r U - \frac{1}{r} U + \frac{1}{2} \partial_\psi \left(U^2 + V^2 + F^2 \right) = -r^2 \partial_\psi P, \tag{3.3.24}$$

which, in view of the second equation in (3.3.23), becomes

$$\partial_r U - \frac{U}{r} = -r^2 \partial_\psi (A + P), \tag{3.3.25}$$

and, using (3.3.19) we see that

$$\partial_\psi p = \frac{1}{r} \partial_r \left(\frac{U}{r} \right), \tag{3.3.26}$$

which then can be used to determine p from knowledge of U. We observe that in
order to have $p = p(\psi)$ a function of ψ alone, from (3.3.26) we have to have

$$U = r^3 M(\psi) + r N(\psi). \tag{3.3.27}$$

for some functions M, N of ψ. Let us denote

$$Q_2(r, \psi) = 2r^2 A(\psi) - F^2(\psi), \tag{3.3.28}$$

$$Q_3(r, \psi) = r^3 M(\psi) + r N(\psi), \tag{3.3.29}$$

and

$$Q_6(r, \psi) = Q_2(r, \psi) - (Q_3(r, \psi))^2 \tag{3.3.30}$$

polynomials of degree 2, 3 and 6 in r with smooth and yet unknown coefficients
depending only on ψ. We note that, in view of (3.3.27),

$$U = Q_3, \tag{3.3.31}$$

and that the second equation in (3.3.23) yields

$$V^2 = Q_6. \tag{3.3.32}$$

Multiplying (3.3.22) by V results in

$$\partial_r Q_6 + Q_3 \partial_\psi Q_6 - 2(\partial_\psi Q_3) Q_6 = 0. \tag{3.3.33}$$

Identifying coefficients in the 9th order polynomial equation (3.3.33) we observe that only odd powers appear, the equations for powers 9 and 7 are tautological, and the remaining three equations become a system of 3 first order ODEs with four unknown functions which is equivalent to the compatibility relation (3.3.22). In order to localize the sought solution u in (r, z) space we need the pressure p to take a value at a point (r_0, z_0) which is strictly separated from all the values it takes on a circle in (r, z) around that point. We seek then conditions which result in a strict local minimum for the function ψ at the chosen point (r_0, z_0), and then a similar behavior for the resulting p. Without loss of generality we may take this local minimum value of ψ to be zero. Because U and V represent derivatives of ψ we are lead to the requirement that the polynomials Q_3 and Q_6 both vanish at the point $(r_0, 0)$ in the (r, ψ) plane, $Q_3(r_0, 0) = 0$ and $Q_2(r_0, 0) = 0$. This results in singular, non-Lipschitz ODEs. They do have nontrivial solutions though, and the consequence given in [16] is

Theorem 3.6 *Let $\ell > 0$, $\tau > 0$ be given. There exists $\epsilon > 0$ and a function $\psi \in C^\infty(B)$, where $B = \{(r, z) \mid |r - \ell|^2 + |z|^2 < \epsilon^2 \ell^2\}$ satisfying $\psi(\ell, 0) = 0$, $\psi > 0$ in B and such that (3.3.18) holds with A, P and F^2 real analytic functions of ψ. The Grad–Shafranov equation (3.3.11) is solved pointwise and has classical solutions in B. The associated velocity u given by the Grad–Shafranov ansatz (3.3.3) is Hölder continuous in B. The Euler equation (3.3.9, 3.3.4) holds weakly in B. The pressure is given by $p = \frac{1}{\ell\tau}\psi$. The vorticity is bounded, $\omega \in L^\infty(B)$ and (3.3.10) holds a.e. in B.*

We note that $F(\psi)$ vanishes like $\sqrt{\psi}$. Therefore, while the ansatz (3.3.3) gives a bounded swirl and a Hölder continuous velocity, the vorticity is not smooth. In fact, in view of (3.3.8) the vorticity equals

$$\omega(r, z) = -F'(\psi)u(r, z) + \text{smooth}. \tag{3.3.34}$$

Thus, $\omega \in L^\infty(B)$, because u vanishes to first order at $(\ell, 0)$, but the r derivative of the z component of vorticity is infinite there.

Once ψ has been constructed so that it has a local minimum at $(\ell, 0)$, then p has also a local minimum there, because, by (3.3.26),

$$p = \frac{\psi}{\ell\tau} \tag{3.3.35}$$

is monotonic in ψ.

Theorem 3.5 holds because the cutoff can be chosen so that the point $(\ell, 0)$ is omitted. By choosing a suitable cutoff function $\phi_\epsilon(p)$, the solution $\tilde{u} = \phi_\epsilon(p)u$ is supported in the region $A = \{(r, z) \mid \frac{1}{2}\ell^2\epsilon^2 < |r - \ell|^2 + |z|^2 < \epsilon^2\ell^2\}$. A consequence of Theorem 3.6 is the following.

Theorem 3.7 *Let $0 < \alpha < 1$. In any domain $\Omega \subset \mathbb{R}^3$ there exist steady solutions of Euler equations belonging to $C^\alpha(\Omega)$ and vanishing at $\partial\Omega$, but such that they do not* belong to $C^\beta(\Omega)$ for $\beta > \alpha$. There exist such solutions which are locally smooth, meaning that for every $x \in \Omega$ there exists a neighborhood of x where the solution is C^∞. For any $\Gamma > 0$, there exist steady solutions u which belong to $L^2(\Omega) \cap C^{\frac{1}{3}}(\Omega)$, vanish at $\partial\Omega$, are locally smooth and have

$$\sup_{x \in \Omega} |\nabla u(x)||u(x)|^2 \geq \Gamma,$$

while the local dissipation vanishes, i.e. $u \cdot \nabla(\frac{|u|^2}{2} + p) = 0$ in the sense of distributions. There exist steady solutions which are locally smooth and whose Lagrangian trajectories have arbitrary linking numbers. For any $0 < \alpha < \beta \leq 1$ there exist weak solutions which are compactly supported in Ω, belong to $C^\alpha(\Omega)$ but not to $C^\beta(\Omega)$.

3.3.3 Steady Multiscale Navier–Stokes Solutions

Proof of Theorem 3.7 is based on a construction which has consequences for the Navier–Stokes equations as well. We describe them here. We take a basic solution of the Euler equations u_B, p_B solving

$$u_B(x) \cdot \nabla u_B(x) + \nabla p_B(x) = 0, \quad \nabla \cdot u_B = 0 \tag{3.3.36}$$

in the unit annulus $A = \{x = (r, z) \mid \frac{1}{2} < |r - 1|^2 + |z|^2 < 1\}$ with

$$u_B \in (C_0^\infty(A))^3, \quad \nabla p_B \in C_0^\infty(A) \tag{3.3.37}$$

constructed by the method of Theorem 3.6. We take an open domain $\Omega \subset \mathbb{R}^3$ and take a sequence of points $x_j \in \Omega$, rotations $R_j \in O(3)$, and numbers $L > 0, T > 0$, $\ell > 0$ $\tau > 0$, with associated length scales

$$\ell_j = L2^{-\ell j} \tag{3.3.38}$$

and time scales

$$\tau_j = T2^{-\tau j}, \tag{3.3.39}$$

for $j = 1, 2, \ldots$, such that functions

$$u_j(x) = \frac{L}{T} 2^{(\tau-\ell)j} R_j u_B \left(2^{\ell j} \frac{R_j^*(x - x_j)}{L} \right) \tag{3.3.40}$$

have disjoint supports

$$A_j = x_j + \ell_j R_j(A) \subset \Omega \tag{3.3.41}$$

in Ω. These are annuli which are rotated, dilated and translated versions of A. Note that the supports of the corresponding pressure gradients

$$\nabla p_j(x) = \frac{L}{T^2} R_j(\nabla p_B) \left(2^{\ell j} \frac{R_j^*(x - x_j)}{L} \right) \tag{3.3.42}$$

are also A_j, and thus disjoint as well, and because of the well known invariance with respect of rotations of the Euler equations we have that

$$u_j \cdot \nabla u_j + \nabla p_j = 0, \qquad \nabla \cdot u_j = 0 \tag{3.3.43}$$

holds in A_j. Let us consider now

$$u(x) = \sum_{j=1}^{N} u_j(x). \tag{3.3.44}$$

Note that $u \in C_0^\infty(\Omega)$, and because the supports of u_j are disjoint, we have

$$\|\partial^\alpha u\|_{L^p(\Omega)} = \left(\sum_{j=1}^{N} 2^{pj[(m-1-\frac{3}{p})\ell + \tau]} \right)^{\frac{1}{p}} L^{1+\frac{3}{p}-m} T^{-1} \|\partial^\alpha u_B\|_{L^p(\Omega)} \tag{3.3.45}$$

for any multiindex α of length $|\alpha| = m \geq 0$. In particular, if we demand that

$$a = \frac{\tau}{\ell} \tag{3.3.46}$$

obeys

$$\frac{3}{2} < a < \frac{5}{2}, \tag{3.3.47}$$

then we have that

$$L^{-3} \|\nabla u\|_{L^2(\Omega)}^2 = \frac{1}{T^2} 2^{N(2\tau - 3\ell)} C_1 \tag{3.3.48}$$

and

$$L^{-3} \|u\|_{L^2(\Omega)}^2 = \frac{L^2}{T^2} C_0. \tag{3.3.49}$$

It is natural to consider the wave number scales

$$k = L^{-1}2^{\ell j}.\tag{3.3.50}$$

The energy spectrum $E(k)$ is by definition the contribution of the kinetic energy at scale k, per unit mass and per scale:

$$E(k) = L^{-3}k^{-1}\|u_j\|^2_{L^2(\Omega)}\tag{3.3.51}$$

so, it follows from our construction of u_j that

$$E(k) = \frac{L^3}{T^2}(kL)^{2a-6}.\tag{3.3.52}$$

The range of scales is limited, the smallest length scale is $L2^{-N\ell}$. If we define a viscosity by

$$\nu = \frac{L^2}{T}2^{-N(2\tau-3\ell)}\tag{3.3.53}$$

then from (3.3.48) we have that

$$\epsilon = \nu L^{-3}\|\nabla u\|^2_{L^2(\Omega)} = \frac{L^2}{T^3}C_1.\tag{3.3.54}$$

Inserting in (3.3.52) we have thus

$$E(k) = C_1^{-\frac{2}{3}}\epsilon^{\frac{2}{3}}L^{\frac{5}{3}}(kL)^{2a-6}.\tag{3.3.55}$$

The Kolmogorov–Obukhov spectrum

$$E(k) = C_K\epsilon^{\frac{2}{3}}k^{-\frac{5}{3}}\tag{3.3.56}$$

is the only spectrum in this family of spectra that does not depend on L. It is obtained at the value

$$a = \frac{13}{6}\tag{3.3.57}$$

which is admissible in view of (3.3.47). If we express the viscosity ν of (3.3.53) in terms of the smallest length scale, (the "dissipation scale") $\ell_d = L2^{-N\ell}$ and in terms of the quantity ϵ of (3.3.54) we obtain

$$\nu = C_1^{-\frac{1}{3}}\epsilon^{\frac{1}{3}}L^{\frac{4}{3}-(2a-3)}(\ell_d)^{2a-3}.\tag{3.3.58}$$

Denoting by $k_d = (\ell_d)^{-1}$ the dissipation wave number scale, (largest wave number scale before exponential decay), we have

$$k_d = C_1^{-\frac{1}{3(2a-3)}} \epsilon^{\frac{1}{3(2a-3)}} L^{\frac{4}{3(2a-3)}-1} \nu^{-\frac{1}{2a-3}}. \tag{3.3.59}$$

Again, the only case which does not depend on L is the Kolmogorov–Obukhov spectrum case $a = \frac{13}{6}$ and in that case we obtain the familiar expression

$$k_d^{-1} = \ell_d = c\nu^{\frac{3}{4}}\epsilon^{-\frac{1}{4}}. \tag{3.3.60}$$

We have proved thus, in particular

Theorem 3.8 *Let Ω be an open set in $\mathbb{R}^3$. There exist smooth stationary solutions of the forced Navier–Stokes equations*

$$u \cdot \nabla u + \nabla p = \nu\Delta u + F, \qquad \nabla \cdot u = 0 \tag{3.3.61}$$

with $u \in C_0^\infty(\Omega)$, $\nabla p \in C_0^\infty(\Omega)$, and such that $\nu\|\nabla u\|_{L^2(\Omega)}^2$ is bounded below uniformly as in (3.3.54) as $\nu \to 0$. There is an inertial range of wave number scales $k \in [k_0, k_d]$ and an exponent $x \in (-3, -1)$, $x = 2a - 6$ with a of (3.3.46), such that the dissipation wave number scale $k_d \sim \nu^{-\frac{1}{x+3}}$ (see (3.3.59)) diverges with $\nu \to 0$ and the energy spectrum $E(k)$ obeys

$$E(k) \sim k^x \tag{3.3.62}$$

(see (3.3.52)) in the inertial range. The force F is smooth, compactly supported, and converges to zero as $\nu \to 0$ in $L^p(\Omega)$ for some p (depending on choice of parameter x) with $1 \le p < 2$.

The proof was given in the computation above, because of the tautology

$$u \cdot \nabla u + \nabla p - \nu\Delta u = F \tag{3.3.63}$$

with

$$F = -\nu\Delta u. \tag{3.3.64}$$

We have

$$x = 2a - 6 \tag{3.3.65}$$

with a given in (3.3.46). The only computation that remains to be shown uses (3.3.45) and (3.3.53), and yields

$$\|F\|_{L^p(\Omega)} \le C_p \frac{L^{1+\frac{3}{p}}}{T^2} 2^{-N\ell(2a-3)} \tag{3.3.66}$$

which follows if $\frac{3}{p} > 1 + a$, or

$$\|F\|_{L^p(\Omega)} \le C_p \frac{L^{1+\frac{3}{p}}}{T^2} 2^{N\ell(4-a-\frac{3}{p})} \tag{3.3.67}$$

if $\frac{3}{p} \in [4 - a, 1 + a]$. For each fixed a, we have $p \in [1, \frac{3}{1+a}]$ when $a \in (\frac{3}{2}, 2]$ and $p \in [1, \frac{3}{4-a}]$ when $a \in [2, \frac{5}{2})$.

References

1. U. Frisch, *Turbulence: The Legacy of A.N. Kolmogorov* (Cambridge University Press, Cambridge, 1995)
2. P. Constantin, Note on loss of regularity for solutions of the 3-D incompressible Euler and related equations. Commun. Math. Phys. **104**(2), 311–326 (1986)
3. N. Masmoudi, Remarks about the inviscid limit of the Navier–Stokes system. Commun. Math. Phys. **270**(3), 777–788 (2007)
4. P. Constantin, J. Wu, Inviscid limit for vortex patches. Nonlinearity **8**(5), 735 (1995)
5. P. Constantin, J. Wu, The inviscid limit for non-smooth vorticity. Indiana Univ. Math. J. **45**(1), 67–81 (1996)
6. P. Constantin, T. Drivas, T. Elgindi, Inviscid limit of vorticity distributions in Yudovich class (2019). arXiv preprint arXiv:1909.04651
7. J. Miller, Statistical mechanics of Euler equations in two dimensions. Phys. Rev. Lett. **65**, 2137–2140 (1990)
8. R. Robert, A maximum-entropy principle for two-dimensional perfect fluid dynamics. J. Stat. Phys. **65**, 531–551 (1991)
9. J. Kelliher, Observations on the vanishing viscosity limit. Trans. Am. Math. Soc. **369**(3), 2003–2027 (2017)
10. H. Bahouri, J.-Y. Chemin, Equations de transport relatives des champs de vecteurs non-lipschitziens et mecanique des fluides. Arch. Rational Mech. Anal. **127**(2), 159–181 (1994)
11. P. Constantin, G. Iyer, A stochastic Lagrangian representation of the three-dimensional incompressible Navier–Stokes equations. Commun. Pure Appl. Math. **61**(3), 330–345 (2008)
12. H. Bahouri, J.-Y. Chemin, R. Danchin, *Fourier Analysis and Nonlinear Partial Differential Equations*, vol. 343 (Springer, Berlin, 2011)
13. E. Stein, *Harmonic Analysis: Real-Variable Methods, Orthogonality, and Oscillatory Integrals* (Princeton University Press, Princeton, 1993)
14. F.-J.L. Nirenberg, On functions of bounded mean oscillation. Commun. Pure Appl. Math., **XIV**, 415–426 (1961)
15. J.-Y. Chemin, A remark on the inviscid limit for two-dimensional incompressible fluids. Commun. Part. Differ. Equ. **21**(11–12), 1771–1779 (1996)
16. P. Constantin, J. La, V. Vicol, Remarks on a paper by Gavrilov: Grad–Shafranov equations, steady solutions of the three-dimensional incompressible Euler equations with com-

pactly supported velocities, and applications. Geom. Funct. Anal. **29**, 1773–1793 (2019). arXiv:1903.11699 [math.AP]

17. A.V. Gavrilov, A steady Euler flow with compact support. Geom. Funct. Anal. **29**(1), 190–197 (2019)

18. D. Chae, P. Constantin, Remarks on a Liouville-type theorem for Beltrami flows. Int. Math. Res. Not. IMRN **20**, 10012–10016 (2015)

19. N. Nadirashvili, Liouville theorem for Beltrami flow. Geom. Funct. Anal. **24**(3), 916–921 (2014)

20. L.E. Fraenkel, M.S. Burger, A global theory of steady vortex rings in an ideal fluid. Acta Math. **132**, 14–51 (1974)

21. H. Grad, H. Rubin, Hydromagnetic equilibria and force-free fields, in *Proceedings of the Second United Nations Conference on the Peaceful Uses of Atomic Energy*, vol. 31 (1958), pp. 190–197

22. V.D. Shafranov, On magnetohydrodynamical equilibrium configurations. Soviet Phys. JETP **6**, 545–554 (1958)

23. T. Buckmaster, V. Vicol, Convex integration and phenomenologies in turbulence (2019). arXiv:1901.09023

24. C. De Lellis, L. Szekelyhdi Jr., High dimensionality and h-principle in PDE. Bull. Am. Math. Soc. **54**(2), 247–282 (2017)

25. Q. Jiu, Z. Xin, Smooth approximations and exact solutions of the 3D steady axisymmetric Euler equations. Commun. Math. Phys. **287**, 323–350 (2009)

26. L.S. Sovolev, The theory of hydromagnetic stability of toroidal plasma configurations. Sov. Phys. JETP **26**(2), 400–407 (1968)

27. V. Yudovich, Nonstationary flow of an ideal incompressible liquid. Zhurn. Vych. Mat. **3**, 1032–1066 (1963). (Russian)

28. L. Onsager, Statistical hydrodynamics. Il Nuovo Cimento (1943–1954) **6**, 279–287 (1949)

29. J. Sommeria, C. Staquet, R. Robert, Final equilibrium state of a two-dimensional shear layer. J. Fluid Mech. **233**, 661–689 (1991)

Chapter 4
Small Scale Creation in Active Scalars

Alexander A. Kiselev

Abstract The focus of the course is on small scale formation in solutions of the incompressible Euler equation of fluid dynamics and associated models. We first review the regularity results and examples of small scale growth in two dimensions. Then we discuss a specific singular scenario for the three-dimensional Euler equation discovered by Hou and Luo, and analyze some associated models. Finally, we will also talk about the surface quasi-geostrophic (SQG) equation, and construct an example of singularity formation in the modified SQG patch solutions as well as examples of unbounded growth of derivatives for the smooth solutions.

4.1 Introduction

In this section we briefly set the stage; for more detailed introduction into our subject here one may consult excellent textbooks [43] or [45]. The incompressible Euler equation of fluid mechanics goes back to 1755 [24]. It appears to be the second PDE ever written down. The equation describes motion of inviscid and incompressible (also called ideal) fluid. The Euler equation is a close relative of the Navier–Stokes equations of fluid mechanics, which came about almost one hundred years later and include viscous effects. One could argue that the Euler equation is less relevant in applications—for example, an observation due to D'Alembert is that there is neither drag nor lift on a body moving in an irrotational ideal fluid. However, Euler equation contains the fluid mechanics nonlinearity, the heart of the Navier–Stokes, and thus for a mathematician it is the first equation to understand. It is also a model of choice in a variety of situations where viscous effects can be ignored. The equation is given by

$$\partial_t u + (u \cdot \nabla)u = \nabla p, \quad \nabla \cdot u = 0. \tag{4.1.1}$$

A. A. Kiselev (✉)
Department of Mathematics, Duke University, Durham, NC, USA
e-mail: kiselev@math.duke.edu

Here $u(x, t)$ is the vector field of the flow, and $p(x, t)$ is pressure. When set in a domain D with boundary, the boundary condition that is natural in many instances is no penetration, $u \cdot n|_{\partial D} = 0$.

The story on global regularity vs. finite time singularity formation question for the Euler equation is very different in two and three dimensions. To see why, it is convenient to look at vorticity $\omega = \mathrm{curl} u$. In the vorticity form, the Euler equation becomes

$$\partial_t \omega + (u \cdot \nabla)\omega = (\omega \cdot \nabla)u, \tag{4.1.2}$$

which is supplemented with the Biot–Savart law which allows to express the velocity u through vorticity. In two dimensions, the Biot–Savart law takes form $u = \nabla^\perp(-\Delta_D)^{-1}\omega$, where Δ_D is the Dirichlet Laplacian on domain D or simply Laplacian on $\mathbb{R}^2$ or $\mathbb{T}^2$, and $\nabla^\perp = (\partial_{x_2}, -\partial_{x_1})$. In three dimensions, $u = \mathrm{curl}(-\Delta)^{-1}\omega$ in $\mathbb{R}^3$ or $\mathbb{T}^3$, while in bounded domains the Biot–Savart law in general takes more complicated form (see [23]).

Coming back to (4.1.2), notice that the term on the right hand side vanishes in two dimensions. This makes equation simpler; notice that for any smooth solution and $p \geq 1$ we have

$$\partial_t \int_D |\omega(x, t)|^p \, dx = p \int_D |\omega|^{p-1} \mathrm{sgn}(\omega)\partial_t \omega \, dx$$

$$= -\int_D u \cdot \nabla |\omega|^p \, dx = \int_D (\nabla \cdot u)|\omega|^p \, dx = 0;$$

where the third step is obtained integrating by parts (using the boundary condition) and the last one by substituting $\nabla \cdot u = 0$. This observation yields many conserved quantities that in three-dimensional case are lacking.

In the next section, we overview existence and uniqueness theory for a fairly general class of solutions to the 2D Euler equation, called Yudovich theory.

4.2 The 2D Euler Equation: A Sketch of Yudovich Theory

In this section, we assume that $D \subset \mathbb{R}^2$ is a smooth bounded domain or that solutions are periodic in space (i.e. $D = \mathbb{T}^2$). More details on the material of this section can be found in [38, 43, 45]. A classical solution of the 2D Euler equation is a C^1 function ω that solves

$$\partial_t \omega + (u \cdot \nabla)\omega = 0, \quad u = \nabla^\perp(-\Delta_D)^{-1}\omega, \quad \omega(x, 0) = \omega_0(x). \tag{4.2.1}$$

It turns out that one can define unique solutions for more general classes of the initial data if one properly modifies the notion of solution.

A key object of the Yudovich theory are particle trajectories $\Phi_t(x)$:

$$\frac{d\Phi_t(x)}{dt} = u(\Phi_t(x), t), \quad \Phi_0(x) = x \qquad (4.2.2)$$

which are defined by incompressible vector field u. If u is smooth, then so is the map $\Phi_t(x)$; this map is also measure preserving since u is divergence free. The map is one-to-one on D by uniqueness of solutions to ordinary differential equations with Lipschitz coefficients. It is not hard to see that it is also onto by solving (4.2.2) backward in time.

A direct calculation shows that ω remains constant on trajectories (again, for smooth solutions), namely, $\frac{d}{dt}\omega(\Phi_t(x), t) = 0$, so

$$\omega(x, t) = \omega_0(\Phi_t^{-1}(x)). \qquad (4.2.3)$$

Next, denote by $G_D(x, y)$ the Green's function of the Dirichlet Laplacian in domain D, so that

$$u(x, t) = \int_D \nabla^\perp G_D(x, y)\omega(y)\, dy. \qquad (4.2.4)$$

A C^1 solution of the Euler equation satisfies the system (4.2.2), (4.2.3), and (4.2.4). We are going to define solutions of low regularity, with ω just in L^∞, by using (4.2.2), (4.2.3), and (4.2.4) instead of (4.2.1). At the heart of the argument are a few simple observations. The first one is a well known estimate of potential theory.

Proposition 4.1 *If $D \subset \mathbb{R}^2$ is a compact domain with a smooth boundary, the Dirichlet Green's function $G_D(x, y)$ has the form*

$$G_D(x, y) = \frac{1}{2\pi} \log |x - y| + h(x, y).$$

Here, for each $y \in D$, $h(x, y)$ is a harmonic function solving

$$\Delta_x h = 0, \quad h|_{x \in \partial D} = -\frac{1}{2\pi} \log |x - y|. \qquad (4.2.5)$$

We have $G_D(x, y) = G_D(y, x)$ for all $(x, y) \in D$, and $G_D(x, y) = 0$ if either x or y belongs to ∂D. In addition, we have the estimates

$$|G_D(x, y)| \le C(D) \left(|\log |x - y|| + 1 \right) \qquad (4.2.6)$$

$$|\nabla G_D(x, y)| \le C(D)|x - y|^{-1}, \qquad (4.2.7)$$

$$|\nabla^2 G_D(x, y)| \le C(D)|x - y|^{-2}. \qquad (4.2.8)$$

The following lemma is a consequence of Proposition 4.1.

Lemma 4.1 *The kernel* $K_D(x, y) = \nabla^{\perp} G_D(x, y)$ *satisfies*

$$\int_D |K_D(x, y) - K_D(x', y)| \, dy \leq C(D)\phi(|x - x'|), \qquad (4.2.9)$$

where

$$\phi(r) = \begin{cases} r(1 - \log r) & r < 1 \\ 1 & r \geq 1, \end{cases} \qquad (4.2.10)$$

with a constant $C(D)$ which depends only on the domain D.

Proof Set $|x - x'| = r > 0$, split the integration into $B_{2r}(x) \cap D$ and its complement, and use estimates of the Proposition 4.1. We leave details to the interested reader, they can also be found in [38, 45].

A result on the regularity of fluid velocity is an immediate consequence.

Corollary 4.1 *Lt D be smooth bounded domain. Suppose the vorticity ω is bounded. Then fluid velocity u satisfies*

$$\|u\|_{L^{\infty}} \leq C(D)\|\omega\|_{L^{\infty}}, \qquad (4.2.11)$$

and

$$|u(x) - u(x')| \leq C\|\omega\|_{L^{\infty}}\phi(|x - x'|), \qquad (4.2.12)$$

with the function $\phi(r)$ defined in (4.2.10).

Proof The estimate (4.2.11) follows from (4.2.7). The proof of (4.2.12) is immediate from Lemma 4.1, as

$$u(x, t) = \int_D K_D(x, y)\omega(y, t)dy.$$

We say that u is log-Lipschitz if it satisfies (4.2.12). A key component of the Yudovich theory is the analysis of the fluid particle trajectories (4.2.2). The classical requirement for uniqueness of solutions to a system of ODE is Lipschitz dependence of coefficients on the unknowns. If the vorticity is bounded, heuristically we expect the velocity u to be one derivative more regular, which would just match the requirement. However, the L^{∞} is the endpoint setting, where we lose a logarithm in regularity, leading to estimates on velocity that are just log-Lipschitz. We will see later that this estimate is sharp and one cannot in general expect better regularity of the velocity corresponding to bounded vorticity. A key observation of the Yudovich theory is that we can still define fluid particle trajectories (4.2.2) uniquely if the velocity u is only log-Lischitz.

The following lemma addresses this question [38, 45].

Lemma 4.2 *Let D be a bounded smooth domain in $\mathbb{R}^d$. Assume that the velocity field $b(x, t)$ satisfies, for all $t \geq 0$:*

$$b \in C([0, \infty) \times \bar{D}), \quad |b(x, t) - b(y, t)| \leq C\phi(|x - y|), \quad b(t, x) \cdot \nu|_{\partial D} = 0. \tag{4.2.13}$$

Here, the function $\phi(r)$ is given by (4.2.10) and ν is the unit normal to ∂D at point x. Then the Cauchy problem in $\bar{D}$

$$\frac{dx}{dt} = b(x, t), \quad x(0) = x_0, \tag{4.2.14}$$

has a unique global solution. Moreover, if $x_0 \notin \partial D$, then $x(t) \notin \partial D$ for all $t \geq 0$. If $x_0 \in \partial D$, then $x(t) \in \partial D$ for all $t \geq 0$.

Note that the log-Lipschitz regularity is essentially border-line: the familiar example of the ODE

$$\dot{x} = x^\beta, \quad x(0) = 0,$$

with $\beta \in (0, 1)$ does not have the uniqueness property: for example, $x(t) \equiv 0$, and

$$x(t) = \frac{t^p}{p^p}, \quad p = \frac{1}{1 - \beta}$$

are both solutions (and in fact one can find infinitely many solutions by separating from zero at an arbitrary time). Thus ODEs with just Hölder (with an exponent smaller than one) coefficients may have more than one solution. Existence of the solutions, on the other hand, does not really require the log-Lipschitz condition: uniform continuity of $b(x, t)$ and at most linear growth as $|x| \to +\infty$ would be sufficient, see e.g. [11] for the Peano existence theorem. We omit the proof of Lemma 4.2, one can check [45] for details.

Now an iterative scheme can be used to construct a weak solution to the 2D Euler equation with L^∞ vorticity, using (4.2.2), (4.2.3), and (4.2.4). We summarize the results of Yudovich theory in the following theorems.

Theorem 4.1 *Fix any $\omega_0 \in L^\infty(D)$. There exists the unique triple $(\omega(x, t), u(x, t), \Phi_t(x))$ satisfying (4.2.2), (4.2.3) and (4.2.4) such that for every $T > 0$ the vorticity ω belongs to $L^\infty([0, T], L^\infty(D))$ and is weak-$*$ continuous in time in L^∞, the fluid velocity $u(t, x)$ is uniformly bounded and log-Lipschitz in x and t, and $\Phi_t \in C^{\alpha(T)}([0, T] \times \bar{D})$ is measure preserving, invertible mapping of $\bar{D}$, satisfying*

$$\frac{d\Phi_t(x)}{dt} = u(\Phi_t(x), t), \quad \Phi_0(x) = x, \tag{4.2.15}$$

$$\omega(x, t) = \omega_0(\Phi_t^{-1}(x)),$$

$$u(x, t) = \int_D K_D(x, y)\omega(y, t)\, dy.$$

Here $\alpha(T) > 0$ and only depends on $\|\omega_0\|_{L^\infty}$ and time T.

A triple (ω, u, Φ_t) satisfying the conditions of Theorem 4.1 is called Yudovich solution to the 2D Euler equation.

If the initial data ω_0 is more regular, this regularity is inherited by the solution.

Theorem 4.2 *Suppose that $\omega_0 \in C^k(\bar{D})$, $k \geq 1$. Then the solution described in Theorem 4.1, satisfies, in addition, the following regularity properties, for each $t \geq 0$:*

$$\omega(t) \in C^k(\bar{D}), \quad \Phi_t(x) \in C^{k,\alpha(t)}(\bar{D}), \text{ and } u \in C^{k,\beta}(\bar{D}),$$

for all $\beta < 1$. In addition, the kth order derivatives of u are log-Lipschitz.

An important example of a Yudovich solution of the 2D Euler equations is the "singular cross" flow, considered by Bahouri and Chemin [1]. We discuss its periodic version here. It corresponds to the vorticity ω_0 which equals to (-1) in the first and third quadrants of the torus $(-\pi, \pi] \times (-\pi, \pi]$, and to $(+1)$ in the other two quadrants:

$$\omega_0(x_1, x_2) = -1 \text{ for } \{0 < x_1, x_2 < \pi\} \text{ and } \{-\pi \leq x_1, x_2 < 0\}, \tag{4.2.16}$$

$$\omega_0(x_1, x_2) = 1 \text{ for } \{0 < x_1 < \pi, -\pi < x_2 < 0\}, \text{ and } \{-\pi < x_1 < 0, 0 < x_2 < \pi\}.$$

We set ω_0 to be equal to zero on the separatrices $x_{1,2} = 0, \pi$. The singular cross has four vortices, one in each quadrant of the torus, and a hyperbolic point at the origin. In fact, ω_0 is a stationary Yudovich solution of the Euler equations. To arrive at this conclusion, the key observation is that ω_0 has the symmetries

$$\omega_0(x_1, x_2) = -\omega_0(-x_1, x_2) = -\omega_0(x_1, -x_2) \tag{4.2.17}$$

on the torus $(-\pi, \pi] \times (-\pi, \pi]$.

Lemma 4.3 *If the initial condition $\omega_0 \in L^\infty$, and satisfies the symmetries (4.2.17), then the Yudovich solution of the 2D Euler equations satisfies the same symmetries for all $t \geq 0$:*

$$\omega(x_1, x_2, t) = -\omega(-x_1, x_2, t) = -\omega(x_1, -x_2, t). \tag{4.2.18}$$

The lemma is proved by checking that if $\omega(x_1, x_2, t)$ is a Yudovich solution of the 2D Euler equation, then so is $-\omega(-x_1, x_2, t)$, and then appealing to the uniqueness property to establish that $\omega(x_1, x_2, t) = -\omega(-x_1, x_2, t)$.

Given odd symmetry and periodicity of $\omega(x, t)$, it is not hard to check that the stream function $\psi := (-\Delta)^{-1}\omega$ is also odd with respect to both variables, with respect to zero as well as $\pm\pi$. Then $u_1 = \partial_{x_2}\psi$ is odd with respect to $x_1 = 0, \pm\pi$ and $u_2 = -\partial_{x_1}\psi$ is odd with respect to $x_2 = 0, \pm\pi$. This implies that the trajectories never leave the quadrants where they originate, and thus by (4.2.3) and (4.2.16) the solution is stationary: $\omega(x, t) \equiv \omega_0(x)$.

The singular cross flow has remarkable properties showing that the estimates on the Yudovich solution of Theorem 4.1 are qualitatively sharp. Namely, the following proposition holds:

Proposition 4.2 *Consider the singular cross solution described above. Then, for small positive x_1, we have*

$$u_1(x_1, 0) = \frac{2}{\pi}x_1 \log x_1 + O(x_1). \tag{4.2.19}$$

The estimate (4.2.19) corresponds to u_1 being just log-Lipschitz near the origin. The proof of (4.2.19) is based on the periodic version of the Biot–Savart law:

Proposition 4.3 *Let $\omega \in L^\infty(\mathbb{T}^2)$ be a mean zero function. Then the vector field*

$$u = \nabla^\perp(-\Delta)^{-1}\omega \tag{4.2.20}$$

is given by

$$u(x) = -\frac{1}{2\pi}\lim_{\gamma\to 0}\int_{\mathbb{R}^2}\frac{(x-y)^\perp}{|x-y|^2}\omega(y)e^{-\gamma|y|^2}\,dy, \tag{4.2.21}$$

where ω has been extended periodically to all $\mathbb{R}^2$.

We leave the proof of this formula as an exercise.

Proof of Proposition 4.2 Given (4.2.21), one can show (4.2.19) by first estimating that the integral over the complement of the central period cell $S = (-\pi, \pi]^2$ contributes regular Lipschitz term to u_1 near the origin. As far as the integral over the central cell goes, let us denote it $u_1^C(x_1, 0)$. It is convenient to go back to representation

$$u_1^C(x_1, 0) = \frac{-\partial_{x_2}}{2\pi}\int_S \log|x-y|\omega_0(y)\,dy = \frac{1}{4\pi}\int_S \partial_{y_2}\log|x-y|^2\omega_0(y)\,dy.$$

Integrating in y_2 over each quadrant and re-grouping the terms, we obtain

$$\frac{1}{2\pi}\left(\int_0^\pi \log\frac{(x_1-y_1)^2}{(x_1-y_1)^2+\pi^2}\,dy_1 - \int_{-\pi}^0 \log\frac{(x_1-y_1)^2}{(x_1-y_1)^2+\pi^2}\,dy_1\right) =$$

$$\frac{1}{\pi}\int_0^\pi \log\frac{x_1-y_1}{x_1+y_1}\,dy_1 + \frac{1}{2\pi}\int_0^\pi \log\frac{(x_1+y_1)^2+\pi^2}{(x_1-y_1)^2+\pi^2}\,dy_1.$$

The last term satisfies

$$\int_0^\pi \log\left(1 + \frac{4x_1 y_1}{(x_1 - y_1)^2 + \pi^2}\right) dy_1 \le Cx_1.$$

Let us split the first term into two parts. First,

$$\int_{2x_1}^\pi \log\left(1 - \frac{2x_1}{y_1 + x_1}\right) dy_1 = -\int_{2x_1}^\pi \frac{2x_1}{y_1 + x_1} dy_1 + O(x_1) = 2x_1 \log x_1 + O(x_1).$$

Second, making the substitution $y_1 = x_1 z$ in the remaining part we obtain

$$\int_0^{2x_1} \log \frac{|x_1 - y_1|}{x_1 + y_1} dy_1 = x_1 \int_0^2 \log \frac{|1 - z|}{|1 + z|} dz = O(x_1).$$

Collecting all the estimates we arrive at (4.2.19).

Since $u_2(x_1, 0) \equiv 0$, a trajectory starting at a point $(x_1^0, 0)$, with $x_1^0 \in (0, \pi)$ is just an interval

$$\Phi_t((x_1^0, 0)) \equiv (x_1(t), 0),$$

moving towards the origin. If x_1^0 is sufficiently small, the component $x_1(t)$ will satisfy

$$x_1'(t) \le x_1(t) \log x_1(t),$$

and so

$$\frac{d}{dt}(\log x_1(t)) \le \log x_1(t),$$

thus

$$\log x_1(t) \le e^t \log x_1^0,$$

and

$$x_1(t) \le x_1(0)^{\exp(t)}. \tag{4.2.22}$$

This estimate has a consequence for the Hölder regularity of the trajectory map. Since the origin is a stationary point of the flow, the inverse flow map $\Phi_t^{-1}(x)$ can be Hölder continuous only with a decaying in time exponent (at most $\sim e^{-t}$). Of course, the direct flow map $\Phi_t(x)$ also has a similar property; to establish it one needs to look at characteristic lines moving along the vertical separatrix. This is exactly the regularity claimed in Theorem 4.1, and thus the Bahouri–Chemin example shows that it cannot be improved.

4.3   The 2D Euler Equation: An Upper Bound on Derivative Growth

More details on material of this section can be found in [16, 36, 38, 43]. We now turn to classical solutions of the 2D Euler equation whose existence and uniqueness are provided by Theorem 4.2. We work in a setting of a compact smooth domain D, but the arguments of this section can be adapted to work on periodic solutions (i.e., $\mathbb{T}^2$) or whole plane $\mathbb{R}^2$. The question that interests us is how quickly can the derivatives of the solutions grow. Such bounds are implicit already in the work of Wolibner [59] and Hölder [31], and have been stated explicitly by Yudovich.

Theorem 4.3 *Assume that $\omega_0 \in C^1(\bar{D})$. Then the gradient of the solution $\omega(x,t)$ satisfies the following bound*

$$\frac{\|\nabla\omega(\cdot,t)\|_{L^\infty}}{\|\omega_0\|_{L^\infty}} \le \left(1 + \frac{\|\nabla\omega_0\|_{L^\infty}}{\|\omega_0\|_{L^\infty}}\right)^{C\exp(\|\omega_0\|_{L^\infty}t)} e^{\exp(C\|\omega_0\|_{L^\infty}t)-1} - 1 \qquad (4.3.1)$$

for all $t \ge 0$.

This upper bound grows at a double exponential rate in time which is extremely fast. A similar double exponential in time upper bound can also be derived for higher order derivatives of vorticity. The occurrence of the double exponential is unusual. Such fast growth, if realized, would pose a formidable challenge in numerical simulations. Let us sketch an argument leading to the estimate (4.3.1). There are three essential ingredients.

1. The kinematics. Using the trajectories Eq. (4.2.2) and some simple estimates, one can derive the following inequality.

$$\exp\left(-\int_0^t \|\nabla u(\cdot,s)\|_{L^\infty}\,ds\right) \le \frac{|\Phi_t(x)-\Phi_t(y)|}{|x-y|} \le \exp\left(\int_0^t \|\nabla u(\cdot,s)\|_{L^\infty}\,ds\right).$$
$$(4.3.2)$$

Since this bound is two sided, it also applies to the inverse map $\Phi_t^{-1}(x)$.

2. The vorticity conservation along trajectories. The formula (4.2.3) implies that

$$\|\nabla\omega(\cdot,t)\|_{L^\infty} \le \|\nabla\omega_0\|_{L^\infty}\sup_{x,y}\frac{|\Phi_t^{-1}(x)-\Phi_t^{-1}(y)|}{|x-y|}. \qquad (4.3.3)$$

3. The Kato inequality.

$$\|\nabla u\|_{L^\infty} \le C(\alpha,D)\|\omega\|_{L^\infty}\left(1 + \log\left(1 + \frac{\|\nabla\omega\|_{L^\infty}}{\|\omega\|_{L^\infty}}\right)\right). \qquad (4.3.4)$$

The way to think about this inequality is as follows. The derivatives of u can be expressed as second order derivatives of the stream function, $\partial^2_{x_i x_j}(-\Delta_D)^{-1}\omega$. Such expressions are called (double) Riesz transforms of ω. These are classical objects in Fourier analysis, and lead to Caldreon–Zygmund operators. Riesz transforms are bounded on L^p, $1 < p < \infty$, but not in L^∞ or L^1 (see e.g. [47]). The structure of the problem, however, requires an L^∞ bound and then we have to pay a logarithm of a higher order norm.

Given (4.3.2), (4.3.3), and (4.3.4), the estimate (4.3.1) follows from some algebraic manipulations and application of Gronwall inequality.

4.4 The 2D Euler Equation: An Example of Double Exponential Growth

More details on material of this section can be found in [36, 38].

A natural question prompted by Theorem 4.3 is whether the double exponential upper bound on growth of derivatives of vorticity is sharp. A variety of examples with some growth in derivatives have been provided by a number of authors. Yudovich [61, 62] built first such examples with growth near the boundary using Lyapunov-type functionals, but without explicit growth rate bounds. Nadirashvili's [48] example is set in an annulus, and is based on using a perturbation of a stable background shear flow. The rate of growth in this example can be shown to be linear in time. Denisov [16] has constructed a periodic solution such that $\|\nabla\omega\|_{L^\infty}$ grows faster than linearly (in a certain average sense). In [17], Denisov has constructed examples with extremely strong (double exponential in time in a certain sense) bursts of growth in derivatives over finite time interval. The idea of Denisov's construction goes back to the singular cross example. Indeed, imagine that a smooth passive scalar $\psi(x, t)$ is advected by the singular cross flow u, that is,

$$\partial_t \psi + (u \cdot \nabla)\psi = 0. \tag{4.4.1}$$

Assume that the initial data ψ_0 is a C_0^∞ bump supported away from the origin and but nonzero on the $x_2 = 0$ separatrix for $x_1 \sim \delta > 0$, δ sufficiently small. Then $\psi(\Phi_t(\delta, 0), t) = \psi_0(\delta) > 0$, while (4.2.22) shows that $\Phi_t(\delta, 0) \leq \delta^{\exp(t)}$. At the same time, $\psi(0, t) = 0$ since the origin is a stationary point of the singular cross flow. Together, these observations imply that

$$\|\nabla\psi(\cdot, t)\|_{L^\infty} \geq \psi_0(\delta)\delta^{-\exp(t)},$$

and is therefore growing at a double exponential rate. The 2D Euler equation for vorticity has the same form as (4.4.1), but of course it is not passive. Changes in ω affect u. Yet the idea of Denisov is to smooth out the singular cross at a very small scale (which depends on how long we would like to control the solution), and to place a perturbation close to the separatrix. In the end, one can mimic the effect of the singular cross flow on passive scalar small scale formation for a finite time, but then control is lost.

The purpose of this section is to present an example where double exponential growth in $\|\nabla\omega\|_{L^\infty}$ is maintained for all times, thus showing that the upper bound of Theorem 4.3 is in general qualitatively sharp [36].

Theorem 4.4 *Consider the two-dimensional Euler equation on a unit disk D. There exist smooth initial data ω_0 with $\|\nabla\omega_0\|_{L^\infty}/\|\omega_0\|_{L^\infty} > 1$ such that the corresponding solution $\omega(x, t)$ satisfies*

$$\frac{\|\nabla\omega(x, t)\|_{L^\infty}}{\|\omega_0\|_{L^\infty}} \geq \left(\frac{\|\nabla\omega_0\|_{L^\infty}}{\|\omega_0\|_{L^\infty}}\right)^{c\exp(c\|\omega_0\|_{L^\infty}t)} \tag{4.4.2}$$

for some $c > 0$ and for all $t \geq 0$.

The example can be extended to any smooth domain with an axis of symmetry [60].

From now on in this section, we will denote by D the unit disk in the plane. It will be convenient for us to take the system of coordinates centered at the lowest point of the disk, so that the center of the disk is at $(0, 1)$. Our initial data $\omega_0(x)$ will be odd with respect to the vertical axis: $\omega_0(x_1, x_2) = -\omega_0(-x_1, x_2)$.

We will take smooth initial data $\omega_0(x)$ so that $\omega_0(x) \leq 0$ for $x_1 > 0$ (and so $\omega_0(x) \geq 0$ for $x_1 < 0$). This configuration makes the origin a hyperbolic fixed point of the flow; in particular, u_1 vanishes on the vertical axis. Let us analyze the Biot–Savart law we have for the disk to gain insight into the structure of the velocity field. The Dirichlet Green's function for the disk is given explicitly by $G_D(x, y) = -\frac{1}{2\pi}(\log|x - y| - \log|x - \bar{y}| - \log|y - e_2|)$, where with our choice of coordinates $\bar{y} = e_2 + (y - e_2)/|y - e_2|^2$, $e_2 = (0, 1)$. Given the symmetry of ω, we have

$$u(x, t) = \nabla^\perp \int_D G_D(x, y)\omega(y, t)\,dy = -\frac{1}{2\pi}\nabla^\perp \int_{D^+} \log\left(\frac{|x - y||\tilde{x} - \bar{y}|}{|x - \bar{y}||\tilde{x} - y|}\right)\omega(y, t)\,dy, \tag{4.4.3}$$

where D^+ is the half disk where $x_1 \geq 0$, and $\tilde{x} = (-x_1, x_2)$. The following Lemma is crucial for the proof of Theorem 4.4. To state it, we need a bit more notation. Let us introduce notation $Q(x_1, x_2)$ for a region that is the intersection of D^+ and the quadrant $x_1 \leq y_1 < \infty$, $x_2 \leq y_2 < \infty$. Given $\pi/2 > \gamma > 0$, denote D_1^γ the intersection of D^+ with a sector $\pi/2 - \gamma \geq \phi \geq 0$, where ϕ is the usual angular variable. Similarly, define D_2^γ the intersection of D^+ with a sector $\pi/2 \geq \phi \geq \gamma$.

Lemma 4.4 *Fix the value of γ, $\pi/2 > \gamma > 0$ (later it will be convenient to take γ sufficiently small, in particular $\gamma < \pi/4$). Suppose that $x \in D_1^\gamma$. Then there exists $\delta > 0$ such that for all $x \in D_1^\gamma$ such that $|x| \leq \delta$ we have*

$$u_1(x_1, x_2, t) = \frac{4}{\pi}x_1 \int_{Q(x_1, x_2)} \frac{y_1 y_2}{|y|^4}\omega(y, t)\,dy_1 dy_2 + x_1 B_1(x_1, x_2, t), \tag{4.4.4}$$

where $\|B_1(\cdot, t)\|_\infty \leq C(\gamma)\|\omega_0\|_{L^\infty}$.

If $x \in D_2^{\gamma}$ is such that $|x| \leq \delta$ then we have

$$u_2(x_1, x_2, t) = -\frac{4}{\pi}x_2 \int_{Q(x_1, x_2)} \frac{y_1 y_2}{|y|^4}\omega(y, t)\, dy_1 dy_2 + x_2 B_2(x_1, x_2, t), \quad (4.4.5)$$

where $\|B_2(\cdot, t)\|_{\infty} \leq C(\gamma)\|\omega_0\|_{L^{\infty}}$.

The proof of the lemma is based on careful analysis of the Biot–Savart law; the details can be found in [36, 38].

It will be convenient to denote

$$\Omega(x_1, x_2, t) = -\frac{4}{\pi}x_2 \int_{Q(x_1, x_2)} \frac{y_1 y_2}{|y|^4}\omega(y, t)\, dy_1 dy_2. \quad (4.4.6)$$

Now let us select the initial data as follows. Fix some small γ, and choose $\delta < 1$ so that the bounds of Lemma 4.4 hold. Note that in what follows, we can always make δ smaller if necessary. Take $\omega_0(x) = -1$ if $x_1 \geq \delta$, odd with respect to x_1, and satisfying $0 \geq \omega_0(x) \geq -1$ for $x \in D^+$.

Lemma 4.5 *Let the initial data ω_0 be as above. Suppose that $|x| \leq \delta$. Then, if δ is sufficiently small, we have*

$$\Omega(x_1, x_2, t) \geq c \log \delta^{-1} \quad (4.4.7)$$

for some universal constant $c > 0$. Here $\Omega(x_1, x_2, t)$ is given by (4.4.6).

Proof We sketch the proof of this estimate leaving detailed computations to the reader. The key observation is that the trajectory map $\Phi_t(x)$ and its inverse are area preserving, while $\omega(x, t) = \omega_0(\Phi_t^{-1}(x))$. For this reason, for all times t, the area of the set $S_t \subset D^+$ where $-1 < \omega(x, t) < 0$ does not exceed 2δ; on the complement of S_t we have $\omega(x, t) = -1$. Given this observation, (4.4.4) and the formula (4.4.6), it is not hard to devise a lower bound estimate that will show (4.4.7). The singularity of the kernel in (4.4.6) would yield $\log|x|^{-1} \gtrsim \log \delta^{-1}$ if integrated against -1 over all $Q(x_1, x_2)$, and removing a set of measure $\leq 2\delta$ from the integration region will preserve the lower bound by $c \log \delta^{-1}$.

Now we will put one more requirement on the initial data. Given $0 < x_1' < x_1'' < 1$, we set

$$O(x_1', x_1'') = \left\{(x_1, x_2) \in D^+,\ x_1' < x_1 < x_1'',\ x_2 < x_1\right\}. \quad (4.4.8)$$

We are going to take a sufficiently small $\epsilon < \delta$ and require in addition that $\omega_0(x) = -1$ for $x \in O(\epsilon^{10}, \epsilon)$. We can find ω_0 satisfying this requirement such that $\|\nabla\omega_0\|_{L^{\infty}} \lesssim \epsilon^{-10}$.

Let us also define, for $0 < x_1 < 1$,

$$\underline{u_1}(x_1, t) = \min_{(x_1, x_2)\in D^+,\, x_2 \leq x_1} u_1(x_1, x_2, t) \quad (4.4.9)$$

and

$$\bar{u}_1(x_1, t) \quad = \quad \max_{(x_1,x_2)\in D^+, \, x_2 \leq x_1} u_1(x_1, x_2, t) . \tag{4.4.10}$$

Since $\omega(x, t)$ and $u(x, t)$ are smooth by Theorem 4.2, these functions are locally Lipschitz in x_1 on $[0, 1)$, with the Lipschitz constants being locally bounded in time. Hence, we can uniquely define $a(t)$ by

$$a' = \bar{u}_1(a, t), \quad a(0) = \epsilon^{10}, \tag{4.4.11}$$

and $b(t)$ by

$$b' = \underline{u}_1(b, t), \quad b(0) = \epsilon . \tag{4.4.12}$$

We set

$$O_t = O(a(t), b(t)) ; \tag{4.4.13}$$

note that O_0 is exactly the set where we set $\omega_0 = -1$ (in addition to the $x_1 \geq \delta$ region). The next key observation is

Lemma 4.6 *We have $\omega(x, t) = -1$ for $x \in O_t$ for all $t \geq 0$.*

Note that for what we know so far, O_t may become empty at some point in time. We will see later that this is not the case.

Proof Here is the sketch of the argument. Due to Lemma 4.5, if δ is chosen sufficiently small, then O_t will lie in the region $x_1 \leq \delta$ for all times and thus the estimates of Lemma 4.4 will continue to apply for all times. The main idea is that given any point $y \in O_t$, we have $y = \Phi_t(x)$ with $x \in O_0$. If we can show that, then the lemma is proved by (4.2.3). Suppose that this is not true, and some trajectory $\Phi_t(x)$ for $x \notin O_0$ ends up inside O_t. This trajectory cannot enter through the boundary ∂D due to the boundary condition. It also could not have entered through the left and right sides of the region O_s due to definitions of a, b, $\bar{u}$ and $\underline{u}$. It remains to consider the diagonal $x_1 = x_2$. However, by Lemma 4.4 and Lemma 4.5, we have

$$\frac{c \log \delta^{-1} - C}{c \log \delta^{-1} + C} \leq \frac{-u_1(x_1, x_1, t)}{u_2(x_1, x_1, t)} \leq \frac{c \log \delta^{-1} + C}{c \log \delta^{-1} - C} .$$

If δ is sufficiently small, this shows that any trajectory on the diagonal part of the boundary of O_t is always moving out of O_t, and hence no trajectory could have entered O_t through the diagonal.

Now we are going to complete the construction of the example.

Proof of Theorem 4.4 To get double exponential growth of the derivatives, we need a genuinely nonlinear argument; it is not sufficient to show that $\Omega(x, t)$ is large for all times as we did in Lemma 4.5. Instead, we will show that $\Omega(x, t)$ grows due to region O_t approaching the origin and remaining sufficiently large. Lemma 4.4 yields

$$\underline{u}_1(b(t), t) \geq -b(t)\,\Omega(b(t), x_2(t), t) - C\,b(t),$$

for some $0 \leq x_2(t) \leq b(t)$ and a constant C that may depend only on γ. A simple calculation shows that, for any $0 \leq x_2 \leq b(t)$ we have

$$\Omega(b(t), x_2, t) \leq \Omega(b(t), b(t), t) + C.$$

Thus, we get

$$\underline{u}_1(b(t), t) \geq -b(t)\,\Omega(b(t), b(t), t) - C\,b(t), \qquad\qquad (4.4.14)$$

with C a new universal constant; below the constant C may change from step to step. In the same vein, for suitable $\widetilde{x}_2(t)$ with $0 \leq \widetilde{x}_2(t) \leq a(t)$, we have

$$\overline{u}_1(a(t), t) \leq -a(t)\,\Omega(a(t), \widetilde{x}_2(t), t) + Ca(t) \leq -a(t)\,\Omega(a(t), 0, t) + Ca(t).$$

A key observation is that

$$\Omega(a(t), 0, t) \geq -\frac{4}{\pi}\int_{O_t}\frac{y_1 y_2}{|y|^4}\omega(t, y)\,dy_1 dy_2 + \Omega(b(t), b(t), t).$$

Since $\omega(y, t) = -1$ on O_t, we have

$$-\int_{O_t}\frac{y_1 y_2}{|y|^4}\omega(t, y)\,dy_1 dy_2 \geq \int_{\pi/8}^{\pi/4}\int_{a(t)/\cos\phi}^{b(t)/\cos\phi}\frac{\sin 2\phi}{2r}\,dr d\phi > \frac{1}{8}(-\log a(t) + \log b(t)).$$

Therefore

$$\overline{u}_1(a(t), t) \leq -a(t)\left(\frac{1}{2\pi}(-\log a(t) + \log b(t)) + \Omega(b(t), b(t), t)\right) + Ca(t).$$
$$(4.4.15)$$

It follows from (4.4.14) and (4.4.15) that

$$\frac{d}{dt}(\log a(t) - \log b(t)) \leq \frac{1}{2\pi}(\log a(t) - \log b(t)) + C. \qquad\qquad (4.4.16)$$

From (4.4.16), the Gronwall lemma leads to

$$\log a(t) - \log b(t) \leq \log \left(a(0)/b(0)\right) \exp(t/2\pi) + C \exp(t/2\pi) \leq (9 \log \epsilon + C) \exp(t/2\pi).$$
(4.4.17)

We should choose our ϵ so that $-\log \epsilon$ is larger than the constant C that appears in (4.4.17). In this case, we obtain from (4.4.17) that

$$\log a(t) \leq 8 \exp(t/2\pi) \log \epsilon,$$

and so

$$a(t) \leq \epsilon^{8 \exp(t/2\pi)}.$$

Note that by the definition of $a(t)$, the first coordinate of the characteristic that originates at the point on ∂D near the origin with $x_1 = \epsilon^{10}$ does not exceed $a(t)$. To arrive at (4.4.2), it remains to note that we chose ω_0 so that $\|\nabla \omega_0\|_{L^\infty} \lesssim \epsilon^{-10}$.

4.5 The Hou-Luo Scenario for the 3D Euler Equation

More information on the material of this section can be found in [42, 43]. The global regularity vs finite time blow up question for smooth solutions of the 3D Euler equation is open. There has been much work on local well-posedness, on conditional regularity criteria, as well as on search for singular scenario. How would finite time blow up manifest itself? In general, any loss of regularity by smooth solution qualifies. However, from many conditional criteria one can infer certain minimal conditions needed for blow up. Perhaps the best known and one of the earliest such conditions was proved by Beale, Kato, and Majda [2, 43]. It states that at the singularity formation time T, one must have

$$\lim_{t \to T} \int_0^t \|\omega(\cdot, s)\|_{L^\infty} \, ds = \infty.$$

A few years ago, Hou and Luo [42] have performed an in-depth numerical simulation, identifying a promising singularity formation scenario. The scenario is axi-symmetric (that is, there is no dependence on the angular variable ϕ in the cylindrical coordinates) and odd with respect to $z = 0$ plane. Very fast vorticity growth is observed at a ring of hyperbolic stagnation points of the flow located on the boundary of a cylinder. In fact, the geometry of the scenario is similar to that of the double exponential growth example for the 2D Euler we discussed in the previous section; the paper [36] has been inspired by the numerical simulations of [42].

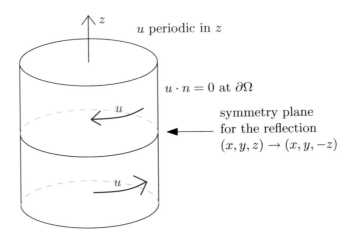

Fig. 4.1 The initial data for Hou-Luo scenario

One of the standard forms of the 3D axi-symmetric Euler equations in the usual cylindrical coordinates (r, ϕ, z) is (see e.g. [43] for more details)

$$\partial_t \left(\frac{\omega^\phi}{r} \right) + u^r \partial_r \left(\frac{\omega^\phi}{r} \right) + u^z \partial_z \left(\frac{\omega^\phi}{r} \right) = \partial_z \left(\frac{(ru^\phi)^2}{r^4} \right) \qquad (4.5.1\text{a})$$

$$\partial_t (ru^\phi) + u^r \partial_r (ru^\phi) + u^z \partial_z (ru^\phi) = 0 , \qquad (4.5.1\text{b})$$

with the understanding that u^r, u^z are given from ω^ϕ via the Biot–Savart law which takes form

$$u^r = -\frac{\partial_z \psi}{r}, \quad u^z = \frac{\partial_r \psi}{r}, \quad L\psi = \frac{\omega^\phi}{r}, \quad L\psi = -\frac{1}{r}\partial_r \left(\frac{1}{r}\partial_r \psi \right) - \frac{1}{r^2}\partial_{zz}^2 \psi .$$

The initial data for the Hou-Luo scenario, shown schematically on Fig. 4.1, has $\omega^\phi = 0$ and only the swirl u^ϕ is non-zero. From (4.5.1), it is clear that the swirl will spontaneously generate toroidal rolls corresponding to non-zero ω^ϕ. These are the so-called "secondary flows", [50]; their effect on river flows was studied by Einstein [19]. Thus the initial condition leads to the (schematic) picture in the xz-plane shown on Fig. 4.2, in which we also indicate the point where a conceivable finite-time singularity (or at least an extremely strong growth of vorticity) is observed numerically. In the three-dimensional picture, the points with very fast growth form a ring on the boundary of the cylinder.

A similar scenario can be considered for the 2D inviscid Boussinesq system in a half-space $\mathbb{R}^+ = \{(x_1, x_2) \in \mathbb{R} \times (0, \infty)\}$ (or in a flat half-cylinder $\mathbf{S}^1 \times (0, \infty)$),

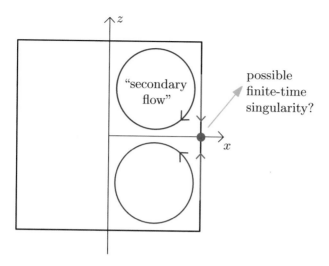

Fig. 4.2 The secondary flows in fixed ϕ section

which we will write in the vorticity form:

$$\partial_t \omega + (u \cdot \nabla)\omega = \partial_{x_1}\theta \tag{4.5.2a}$$

$$\partial_t \theta + (u \cdot \nabla)\theta = 0 . \tag{4.5.2b}$$

Here $u = (u_1, u_2)$ is obtained from ω by the usual Biot–Savart law $u = \nabla^{\perp}(-\Delta)^{-1}\omega$, with appropriate boundary conditions on Δ, and θ represents fluid temperature or density.

It is well-known (see [43]) that this system has properties similar to the 3D axisymmetric Euler (4.5.1), at least away from the symmetry axis. Indeed, comparing (4.5.1) with (4.5.2), we see that θ essentially plays the role of the square of the swirl component ru^{ϕ} of the velocity field u, and ω replaces ω^{ϕ}/r. The real difference between the two systems only emerges near the axis of rotation, where factors of r can conceivably change the nature of dynamics. For the purpose of comparison with the axi-symmetric flow, the last picture should be rotated by $\pi/2$, after which it resembles the picture relevant for (4.5.2), see Figs. 4.2 and 4.3. The system (4.5.2) has an advantage of being simpler looking and easier to think about while very likely preserving all the essential features. Of course, the question of global regularity vs finite time blow up is also open for the 2D inviscid Boussinesq; in fact it appears on the list of "eleven great problems of mathematical hydrodynamics" by Yudovich [63].

In both the 3D axi-symmetric Euler case and in the 2D Boussinesq system case the best chance for possible singularity formation seems to be at the points of symmetry located at the boundary, which numerical simulations suggest are fixed hyperbolic points of the flow. So far there is no proof of singularity formation for smooth solutions in the Hou-Luo scenario, but a number of models have been

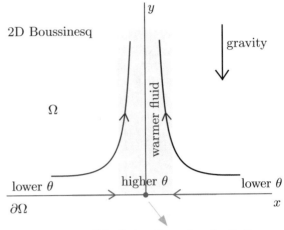

Fig. 4.3 The 2D Boussineq singularity scenario

possible finite-time singularity?

proposed (e.g. [9, 10, 18, 30, 32, 40, 41]). All these models suggest finite time blow up. In the next section, we will take a look at one of the simpler ones that provide a qualitative insight into the nature of possible nonlinear feedback loop leading to singularity.

In addition, there is recent and very interesting work by Elgindi and Elgindi and Jeong [20–22] and Chen and Hou [7] where the initial data is taken to be singular, to a varying degree, in a scenario very similar to that of Hou-Luo and involving stationary hyperbolic points of the flow. Then finite time blow up is shown in a sense of the solution becoming more singular than the initial data. In [20, 21] the domain needs to have a corner (for the 2D Boussinesq, and a wedge for the 3D Euler), and the initial vorticty is just L^∞, but becomes unbounded in finite time. The work [22] is set in the whole space, and the initial vorticity is Hölder continuous, leading to a solution satisfying

$$\lim_{t \to 1} \int_0^t \|\omega(\cdot, t)\|_{L^\infty} \, dt = \infty.$$

The paper [7] builds on [22] and [8] and proves finite time blow up for solutions with C^α initial vorticity and $C^{1,\alpha}$ density (respectively, swirl) in the Hou-Luo scenario.

4.6   Singularities and Turbulence

In this section, we take a step back to look at the big picture. The material in this section is motivational and largely informal. Most of the statements and problems mentioned here are either heuristic or remain wide open. One of the incentives in trying to understand and describe small scale and possible singularity formation in solutions to equations of fluid mechanics is the connection to turbulence. Turbulence is familiar to all of us from bumpy airplane ride or fluctuations we feel in strong wind. It is a ubiquitous feature of intense fluid motion, and plays a role in a wide range of processes in nature: drag effects for cars and airplanes, efficiency of internal combustion engines, mixing crucial for survival of marine animals, or even evolution of temperature inside Earth [57]! Nevertheless, turbulence remains relatively poorly understood: Richard Feynman has called it the greatest unsolved problem of classical physics 50 years ago, and not much has changed since. This situation is not for the lack of trying: a number of heuristic theories have been proposed by Prandtl, von Kármán, Richardson, Taylor, Heisenberg, Kolmogorov, Onsager, Kraichnan, and others, see e.g. [25] for review. These phenomenological theories have been quite successful in predicting some of the properties of turbulent flows, but deeper understanding and in particular rigorous connection to the partial differential equations of fluid mechanics have not been established. Among these predictions, the one most consistent with experiments and reported at least as early as Dryden's wind tunnel experiments in 1943, is Kolmogorov's "zeroth law of turbulence" which postulates *anomalous dissipation of energy*, that is, non-vanishing of the rate of dissipation of kinetic energy of turbulent fluctuations per unit mass, in the limit of zero viscosity. Let u^ν be solutions of the 3D Navier–Stokes equations with viscosity ν (in the non-dimensional form, ν is equal to the inverse of the Reynolds number, $Re = \frac{UL}{\sigma}$, where σ is the actual viscosity and U, L typical velocity and length scales):

$$\partial_t u^\nu + (u^\nu \cdot \nabla)u^\nu - \nu \Delta u^\nu = \nabla p^\nu + f, \quad \nabla \cdot u^\nu = 0,$$

where f is some spatially regular forcing. Notice that $\nu \int |\nabla u^\nu|^2 \, dx$ is just an instantaneous rate of energy dissipation. Then the zeroth law states that

$$\lim_{\nu \to 0} \nu \langle |\nabla u^\nu(x, t)|^2 \rangle > 0, \qquad (4.6.1)$$

where $\langle \cdot \rangle$ represents a suitable ensemble or space-time average. On the other hand, one may expect that as $\nu \to 0$, u^ν converges to the solution u of the incompressible 3D Euler equation. This is far from clear, especially when there are boundaries. But if we accept this as a reasonable assumption, and the limiting solution is smooth, then the limit of the enstrophy $\int |\nabla u^\nu|^2 \, dx$ should equal $\int |\nabla u|^2 \, dx < \infty$, and so (4.6.1) cannot hold. Therefore, if the zeroth law of turbulence is mathematically valid, it is intimately linked with singularity formation in the solutions of the Euler (and possibly Navier–Stokes) equation. In fact, the zeroth law has been rigorously proved in just one PDE setting: for stochastically forced Burgers equation [58]. In this case, the underlying singularities are well understood shocks, and this is a key aspect that makes the problem approachable.

The set up of the Hou-Luo scenario seems to be rather special, but there are indications that hyperbolic point geometry plays an important role in extreme dissipation regions in turbulent fluid flows. In a paper [52], the authors describe a physical experiment where they study turbulence. The geometry of the experiment is very similar to the Hou-Luo setting: the flows are confined to a cylinder and are statistically axi-symmetric (of course, in any real flow the exact axial symmetry breaks down spontaneously). Using particle image velocimetry techniques, they are able to capture the structure of fluid flow in the regions of "extreme inertial dissipation events". These are regions where the dissipation rate is anomalously large, essentially the regions that would facilitate the zeroth law as viscosity decreases. There are four different geometric scenarios documented in [52], and by far the most common is the one where vorticity growth happens at the front between colliding masses of fluid with different directions of the velocity and hyperbolic point geometry. In this sense, the Hou-Luo scenario may be thought of as an idealized blueprint of what appears to be a common small scale formation mechanism.

In recent years, there have been other works on singularity formation in equations of fluid mechanics, in particular by Tao [55, 56]. In these papers, finite time blow up is proved for modified Euler and Navier–Stokes equations; the modifications involve suitable averaging or changes in coupling of the Fourier modes. The philosophy of these examples is different from Hou-Luo scenario and involves instead a self-similar picture of energy transition to smaller and smaller scales. Inspired by these ideas, the paper [4] has outlined a specific physical mechanism that could be behind such self-similar cascades: repeated flattening of vortex tubes into sheets and breakup of the sheets into tubes. There are some numerical simulations, however (see e.g. [46] where further references can be found), that suggest that this process is quite complex and it is not clear whether some sort of approximate self-similarity can be traced over smaller and smaller scales.

4.7   One-Dimensional Models of the Hou-Luo Scenario

More details on the results discussed in this section can be found in [9, 10, 42]. A one-dimensional model of the Hou-Luo scenario was formulated already in [42]. This model is given by

$$\partial_t \omega + u \partial_x \omega = \partial_x \theta,$$

$$\partial_t \theta + u \partial_x \theta = 0, \quad u_x = H\omega. \tag{4.7.1}$$

Here H is the Hilbert transform, and the setting can be either periodic or the entire axis with some decay of the initial data. The model (4.7.1) can be thought of as an effective equation on the $x_2 = 0$ axis in the Boussinesq case (see (4.5.2) and Fig. 4.3) or on the boundary of the cylinder in the 3D axi-symmetric Euler case. The model can be derived from the full equations under assumption that $\omega(x, t)$ is concentrated in a boundary layer of width a near $x_2 = 0$ axis and is independent of x_2, that is $\omega(x_1, x_2, t) = \omega(x_1, t)\chi_{[0,a]}(x_2)$. Such assumption allows to close the system and reduces the half-plane Biot–Savart law to $u_x = H\omega$ in the main order; the parameter a enters into the additional term that is non-singular and is dropped from (4.7.1). See [42], [10] for more details. We will call the system (4.7.1) the Hou-Luo model.

The Hou-Luo model is still fully nonlocal. A further simplification was proposed in [9], where the Biot–Savart law has been replaced with

$$u(x, t) = -x \int_x^1 \frac{\omega(y, t)}{y} \, dy. \tag{4.7.2}$$

Here the most natural setting is on an interval $[0, 1]$ with smooth initial data supported away from the endpoints. For simplicity we set the Biot–Savart law so that blow up at zero happens for positive vorticity. The law (4.7.2) is motivated by the velocity representation in Lemma 4.4 above, as it is the simplest one-dimensional analog of such representation. This law is "almost local"—if one divides u by x and differentiates, one gets local expression. We will call the model (4.7.2) the CKY model. The law (4.7.2) models the situation where ω is odd in x_1, and with this additional assumption one can show that it is not too different from the relation $u_x = H\omega$ in the Hou-Luo model.

Both Hou-Luo and CKY models are locally well-posed in reasonable spaces, and in both cases possibility of finite time blow up has been proved in [9] and [10] respectively. Here we sketch the arguments showing singularity formation in the CKY model [9]. In this section, let us denote

$$\Omega(x, t) = \int_x^1 \frac{\omega(y, t)}{y} \, dy. \tag{4.7.3}$$

The first step is the following

Lemma 4.7 *Along the trajectories* $\Phi_t(x)$, *we have*

$$\frac{d}{dt} \Omega(\Phi_t(x), t) = \int_{\Phi_t(x)}^1 \frac{\omega(y, t)^2}{y} dy + \int_{\Phi_t(x)}^1 \frac{\partial_x \theta(y, t)}{y} dy. \tag{4.7.4}$$

The proof of this lemma is a direct computation taking advantage of (4.7.1), (4.7.2), and integration by parts. A key observation is the positivity of the first term in the right hand side of (4.7.4).

Consider now $\psi(x, t) := \log \Phi_t(x)^{-1}$. From (4.7.2), we have that

$$\partial_t \psi(x, t) = \Omega(\Phi_t(x), t). \tag{4.7.5}$$

On the other hand, from Lemma 4.7, we have that

$$\frac{d}{dt} \Omega(\Phi_t(x), t) \geq \int_{\Phi_t(x)}^1 \frac{\partial_x \theta(y, t)}{y} dy. \tag{4.7.6}$$

Let us trace a trajectory that originates at some point in the support of θ where $\partial_x\theta$ is not zero. As the flow pushes the vorticity support towards the origin, we expect that the front of the graph of θ will become very steep. Then morally, we can think of the integral on the right hand side of (4.7.6) as

$$\int_{\Phi_t(x)}^{1} \frac{\partial_x\theta(y,t)}{y}dy \sim \Phi_t(x)^{-1} = e^{\psi(x,t)},$$

as most of the variation of θ will be supported close to $\Phi_t(x)$. If we accept this heuristic argument for a moment, we get a system of differential inequalities

$$\partial_t\psi(x,t) = \Omega(\Phi_t(x),t), \quad \partial_t\Omega(\Phi_t(x),t) \gtrsim e^{\psi(x,t)}$$

for which it is not hard to prove finite time blow up. Such blow up corresponds to a trajectory carrying some positive value of the density arriving at the origin at a finite time $T < \infty$, which one can show only happens if $\int_0^T \|\omega(\cdot,t)\|_{L^\infty} dt = \infty$.

A more careful argument to establish blow up uses a cascade of trajectories corresponding to a sequence of initial points with larger and larger values of θ_0, and iterative estimates. The details can be found in [9].

4.8 The Modified SQG Equation: Singularity Formation in Patches

More details regarding material of this section can be found in [33, 37, 39]. Note that the 2D Euler equation is just the 2D inviscid Boussinesq system with $\theta \equiv 0$. It is tempting to try to extend the insight and techniques used in the proof of Theorem 4.4 to analysis of the Hou-Luo scenario. There are several issues that arise in such an attempt. Perhaps the most significant one is that the vorticity may no longer be bounded, so the result of Lemma 4.4 giving fairly precise control over fluid velocity near the origin is not available. The kernel in the Biot–Savart law is not sign definite, and growth of vorticity may conceivably make the contributions of the regions that in Lemma 4.4 end up in the regular Lipschitz term no longer relatively small. This might destroy the hyperbolic structure of the flow and sabotage singularity formation. The available evidence suggests that singularity formation likely holds—the numerical computations of Hou and Luo are very detailed and precise, all the models considered so far suggest blow up, and so does the work of Elgindi and Jeong [20–22]. Yet there are counter arguments to all these points. Any numerical simulation has a limit; the models of the Hou-Luo scenario make simplifying assumptions on the Biot–Savart law and other aspects of the problem; and the works on rough initial data do not apply to smooth initial data.

In this section, we discuss a blow up example in a different setting, where nevertheless some of the technical issues are similar. In particular, we will see that the term pushing towards singularity has the same order as the term opposing it (in contrast to Lemma 4.4 where we could isolate a relatively simple dominant main term). This setting is patch solutions to modified surface quasi-geostrophic (SQG) equation.

The SQG equation is similar to the 2D Euler equation in vorticity form, but is more singular:

$$\partial_t \omega + (u \cdot \nabla)\omega = 0, \quad u = \nabla^\perp(-\Delta)^{-1+\alpha}\omega, \quad \alpha = 1/2, \quad \omega(x, 0) = \omega_0(x).$$
$$(4.8.1)$$

The value $\alpha = 0$ in (4.8.1) corresponds to the 2D Euler equation, while $0 < \alpha < \frac{1}{2}$ is the so-called modified SQG range. The SQG and modified SQG equations come from atmospheric science. They model evolution of temperature near the surface of a planet and can be derived by formal asymptotic analysis from a larger system of rotating 3D Navier–Stokes equations coupled with temperature equation through buoyancy force [29, 38, 44, 49]. In mathematical literature, the SQG equation was first considered by Constantin, Majda and Tabak [12], where a parallel between the structure of the SQG equation and the 3D Euler equation was drawn. A singularity formation scenario, a closing front, has been proposed in [12], but it was later proved to be impossible under certain additional assumptions in [13, 14]. The SQG and modified SQG equations are perhaps simplest looking equations of fluid mechanics for which the question of global regularity vs finite time blow up remains open.

Equation (4.8.1) can be considered with smooth initial data, but another important class of initial data is patches, where $\theta_0(x)$ equals linear combination of characteristic functions of some disjoint domains $\Omega_j(0)$. The resulting evolution yields time dependent regions $\Omega_j(t)$. The regularity question in this context addresses the regularity class of the boundaries $\partial\Omega_j(t)$ and lack of self-intersection or contact between different components $\Omega_j(t)$. Existence and uniqueness of patch solution for 2D Euler equation follows from Yudovich theory. The global regularity question has been settled affirmatively by Chemin [6] (Bertozzi and Constantin [3] provided a different proof). For the SQG and modified SQG equations patch dynamics is harder to set up. Local well-posedness has been shown by Rodrigo in C^∞ class [51] and by Gancedo in Sobolev spaces [26] in the whole plane setting. Numerical simulations by Cordoba et al. [15] and by Dritschel and Scott [53, 54] suggest that finite time singularities are possible. There are different scenarios involving boundaries touching and forming corners [15, 54] and self-similar cascade of filaments [53], but rigorous understanding of this phenomena remains missing.

We will discuss modified SQG and Euler patches in a half-plane. The Bio-Savart law for the patch evolution on the half-plane $D := \mathbb{R} \times \mathbb{R}^+$ is

$$u = \nabla^\perp(-\Delta_D)^{-1+\alpha}\omega,$$

with Δ_D being the Dirichlet Laplacian on D, which can also be written as

$$u(x, t) := \int_D \left(\frac{(x - y)^\perp}{|x - y|^{2+2\alpha}} - \frac{(x - \bar{y})^\perp}{|x - \bar{y}|^{2+2\alpha}} \right) \omega(y, t) dy. \tag{4.8.2}$$

Note that u is divergence free and tangential to the boundary. A traditional approach to the 2D Euler ($\alpha = 0$) vortex patch evolution, going back to Yudovich (see [45] for an exposition) is via the corresponding flow map. The active scalar ω is advected by u from (4.8.2) via

$$\omega(x, t) = \omega \left(\Phi_t^{-1}(x), 0 \right), \tag{4.8.3}$$

where

$$\frac{d}{dt} \Phi_t(x) = u \left(\Phi_t(x), t \right) \quad \text{and} \quad \Phi_0(x) = x. \tag{4.8.4}$$

The initial condition ω_0 for (4.8.2)–(4.8.4) is patch-like,

$$\omega_0 = \sum_{k=1}^{N} \theta_k \chi_{\Omega_k(0)}, \tag{4.8.5}$$

with $\theta_1, \ldots, \theta_N \neq 0$ and $\Omega_1(0), \ldots, \Omega_N(0) \subseteq D$ bounded open sets, whose closures $\overline{\Omega_k(0)}$ are pairwise disjoint and whose boundaries $\partial \Omega_k(0)$ are simple closed curves of given regularity.

One reason the Yudovich theory works for the 2D Euler equations is that for ω which is (uniformly in time) in $L^1 \cap L^\infty$, the velocity field u given by (4.8.2) with $\alpha = 0$ is log-Lipschitz in space, and the flow map Φ_t is everywhere well-defined as discussed in Sect. 4.2. But when ω is a patch solution and $\alpha > 0$, the flow u from (4.8.2) is smooth away from the patch boundaries $\partial \Omega_k(t)$ but is only $1 - \alpha$ Hölder continuous near $\partial \Omega_k(t)$, which is exactly where one needs to use the flow map. This creates significant technical difficulties in proving local well-posedness of patch evolution in some reasonable functional space. A naive intuition on why patch evolution can be locally well-posed for $\alpha > 0$ without boundaries is that one can show that the below-Lipschitz loss of regularity only affects the tangential component of the fluid velocity at patch boundary. The normal to patch component, that intuitively should determine the evolution of the patch, retains stronger regularity.

In presence of boundaries, the problem is even harder. Intuitively, one reason for the difficulties can be explained as follows. In the simplest case of the half-plane the reflection principle implies that the boundary can be replaced by a reflected patch (or patches) of the opposite sign. If the patch is touching the boundary, then the reflected and original patch are touching each other, and the low regularity tangential component of the velocity field generated by the reflected patch has strong influence on the boundary of the original patch near touch points. Even in the 2D Euler case, the global regularity for patches in general domains with boundaries remained open until very recently [34]. In the half-plane, a global regularity result has been established earlier in [37]:

Theorem 4.5 *Let $\alpha = 0$ and $\gamma \in (0, 1]$. Then for each $C^{1,\gamma}$ patch-like initial data ω_0, there exists a unique global $C^{1,\gamma}$ patch solution ω to (4.8.3), (4.8.2), and (4.8.4) with $\omega(\cdot, 0) = \omega_0$.*

In the case $\alpha > 0$ with boundary, even local well-posedness results are highly non-trivial. The following result has been proved in [39] for the half-plane.

Theorem 4.6 *If $\alpha \in (0, \frac{1}{24})$, then for each H^3 patch-like initial data ω_0, there exists a unique local H^3 patch solution ω with $\omega(\cdot, 0) = \omega_0$. Moreover, if the maximal time T_ω of existence of ω is finite, then at T_ω a singularity forms: either two patches touch, or a patch boundary touches itself or loses H^3 regularity.*

On the other hand, in [37], it was proved that for any $\frac{1}{24} > \alpha > 0$, there exist patch-like initial data leading to finite time blow up.

Theorem 4.7 *Let $\alpha \in (0, \frac{1}{24})$. Then there are H^3 patch-like initial data ω_0 for which the unique local H^3 patch solution ω with $\omega(\cdot, 0) = \omega_0$ becomes singular in finite time (i.e., its maximal time of existence T_ω is finite).*

Together, Theorems 4.5 and 4.7 give rigorous meaning to calling the 2D Euler equation critical. In the half-plane patch framework $\alpha = 0$ is the exact threshold for phase transition from global regularity to possibility of finite time blow up.

Recently, the local well-posedness result of Theorem 4.6 and the finite time blow up example of Theorem 4.7 have been extended to $0 < \alpha < 1/3$ for H^2 patches in [27].

In what follows, we will sketch the proof of the blow up Theorem 4.7. Full details can be found in [37]. Let us describe the initial data set up. Denote $\Omega_1 := (\epsilon, 4) \times (0, 4)$, $\Omega_2 := (2\epsilon, 3) \times (0, 3)$, and let $\Omega_0 \subseteq D^+ \equiv \mathbb{R}^+ \times \mathbb{R}^+$ be an open set whose boundary is a smooth simple closed curve and which satisfies $\Omega_2 \subseteq \Omega_0 \subseteq \Omega_1$. Here ϵ is a small parameter depending on α that will be chosen later.

Let $\omega(x, t)$ be the unique H^3 patch solution corresponding to the initial data

$$\omega(x, 0) := \chi_{\Omega_0}(x) - \chi_{\widetilde{\Omega}_0}(x) \tag{4.8.6}$$

with maximal time of existence $T_\omega > 0$. Here, $\widetilde{\Omega}_0$ is the reflection of Ω_0 with respect to the x_2-axis. Then

$$\omega(x, t) = \chi_{\Omega(t)}(x) - \chi_{\widetilde{\Omega}(t)}(x) \tag{4.8.7}$$

for $t \in [0, T_\omega)$, with $\Omega(t) := \Phi_t(\Omega_0)$. It can be seen from (4.8.2) that the rightmost point of the left patch on the x_1-axis and the leftmost point of the right patch on the x_1-axis will move toward each other. In the case of the 2D Euler equations

Fig. 4.4 The domains $\Omega_1, \Omega_2, \Omega_0$, and $K(0)$ (with $\omega_0 = \chi_{\Omega_0} - \chi_{\widetilde{\Omega}_0}$)

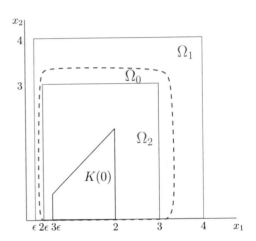

$\alpha = 0$, Theorem 4.5 shows that the two points never reach the origin. When $\alpha > 0$ is small, however, it is possible to control the evolution sufficiently well to show that—unless the solution develops another singularity earlier—both points will reach the origin in a finite time. The argument yielding such control is fairly subtle, and the estimates only work for sufficiently small α, even though one would expect singularity formation to persist for more singular equations. This situation is not uncommon in the field: there is plenty of examples with the infinite in time growth of derivatives for the smooth solutions of 2D Euler equation, while until very recently none were available for the more singular SQG equation—this will be discussed in the Sect. 4.9.

To show finite time blow up, we will deploy a barrier argument. Define

$$K(t) := \{x \in D^+ : x_1 \in (X(t), 2) \text{ and } x_2 \in (0, x_1)\} \tag{4.8.8}$$

for $t \in [0, T]$, with $X(0) = 3\epsilon$. Clearly, $K(0) \subset \Omega(0)$. Set the evolution of the barrier by

$$X'(t) = -\frac{1}{100\alpha} X(t)^{1-2\alpha}. \tag{4.8.9}$$

Then $X(T) = 0$ for $T = 50(3\epsilon)^{2\alpha}$. So if we can show that $K(t)$ stays inside $\Omega(t)$ while the patch solution stays regular, then we obtain that singularity must form by time T : the different patch components will touch at the origin by this time unless regularity is lost before that. See Fig. 4.4 for the set up illustration.

The key step in the proof involves estimates of the velocity near origin. In particular, u_1 needs to be sufficiently negative to exceed the barrier speed (4.8.9); u_2 needs to be sufficiently positive in order to ensure that $\Omega(t)$ cannot cross the barrier along its diagonal part. Note that it suffices to consider the part of the barrier that is very close to the origin, on the order $\sim\epsilon^{2\alpha}$. Indeed, the time T of barrier arrival at the origin has this order, and the fluid velocity satisfies uniform L^∞ bound that follows by a simple estimate which uses only $\alpha < 1/2$. Thus the patch $\Omega(t)$ has no time to reach more distant boundary points of the barrier before formation of singularity.

Let us focus on the estimates for u_1. For $y = (y_1, y_2) \in \bar{D}^+ = \mathbb{R}^+ \times \mathbb{R}^+$, we denote $\bar{y} := (y_1, -y_2)$ and $\tilde{y} := (-y_1, y_2)$. Due to odd symmetry, (4.8.2) becomes (we drop t from the notation in this sub-section)

$$u_1(x) = -\int_{D^+} K_1(x, y)\omega(y)dy, \qquad (4.8.10)$$

where

$$K_1(x, y) = \underbrace{\frac{y_2 - x_2}{|x - y|^{2+2\alpha}}}_{K_{11}(x,y)} - \underbrace{\frac{y_2 - x_2}{|x - \tilde{y}|^{2+2\alpha}}}_{K_{12}(x,y)} - \underbrace{\frac{y_2 + x_2}{|x + y|^{2+2\alpha}}}_{K_{13}(x,y)} + \underbrace{\frac{y_2 + x_2}{|x - \bar{y}|^{2+2\alpha}}}_{K_{14}(x,y)}.$$

$$(4.8.11)$$

Analyzing (4.8.11), it is not hard to see that we can split the region of integration in the Biot–Savart law according to whether it helps or opposes the bounds we seek. Define

$$u_1^{bad}(x) := -\int_{\mathbb{R}^+ \times (0,x_2)} K_1(x, y)\omega(y)dy \text{ and } u_1^{good}(x) := -\int_{\mathbb{R}^+ \times (x_2,\infty)} K_1(x, y)\omega(y)dy.$$

The following two lemmas contain the key estimates.

Lemma 4.8 (Bad Part) *Let $\alpha \in (0, \frac{1}{2})$ and assume that ω is odd in x_1 and $0 \le \omega \le 1$ on D^+. If $x \in \overline{D^+}$ and $x_2 \le x_1$, then*

$$u_1^{bad}(x) \le \frac{1}{\alpha}\left(\frac{1}{1 - 2\alpha} - 2^{-\alpha}\right)x_1^{1-2\alpha}. \qquad (4.8.12)$$

The proof of this lemma uses (4.8.11) and after cancellations leads to the bound

$$u_1^{bad}(x) \le -\int_{(0,2x_1) \times (0,x_2)} \frac{y_2 - x_2}{|x - y|^{2+2\alpha}}dy, \qquad (4.8.13)$$

which gives (4.8.12)

In the estimate of the good part, we need to use a lower bound on ω that will be provided by the barrier. Define

$$A(x) := \{y : y_1 \in (x_1, x_1 + 1) \text{ and } y_2 \in (x_2, x_2 + y_1 - x_1)\}. \qquad (4.8.14)$$

Lemma 4.9 (Good Part) *Let $\alpha \in (0, \frac{1}{2})$ and assume that ω is odd in x_1 and for some $x \in \overline{D^+}$ we have $\omega \geq \chi_{A(x)}$ on D^+, with $A(x)$ from (4.8.14). There exists $\delta_\alpha \in (0, 1)$, depending only on α, such that the following holds. If $x_1 \leq \delta_\alpha$, then*

$$u_1^{good}(x) \leq -\frac{1}{6 \cdot 20^\alpha \alpha} x_1^{1-2\alpha}.$$

Here analysis of (4.8.11) leads to

$$u_1^{good}(x) \leq -\underbrace{\int_{A_1} \frac{y_2 - x_2}{|x - y|^{2+2\alpha}} dy}_{T_1} + \underbrace{\int_{A_2} \frac{y_2 - x_2}{|x - y|^{2+2\alpha}} dy}_{T_2},$$

with the domains

$$A_1 := \{y : y_2 \in (x_2, x_2 + 1) \text{ and } y_1 \in (x_1 + y_2 - x_2, 3x_1 + y_2 - x_2)\},$$

$$A_2 := (x_1 + 1, 3x_1 + 1) \times (x_2, x_2 + 1).$$

The term T_2 can be estimated by Cx_1, since the region of integration A_2 lies at a distance ~ 1 from the singularity. A relatively direct estimate of the term T_1 leads to the result of the Lemma.

A distinctive feature of the problem is that estimates for the "bad" and "good" terms appearing in Lemmas 4.8 and 4.9 have the same order of magnitude $x_1^{1-2\alpha}$. This is unlike the 2D Euler double exponential growth construction, where we were able to isolate the main term. To understand the balance in the estimates for the "bad" and "good" terms, note that the "bad" term estimate comes from integration of the Biot–Savart kernel over rectangle $(0, 2x_1) \times (0, x_2)$, while the good term estimate from integration of the same kernel over the region A_1 above. When α is close to zero, the kernel is longer range, and the more extended nature of the region A_1 makes the "good" term dominate. In particular, the coefficient $\frac{1}{\alpha} \left(\frac{1}{1-2\alpha} - 2^{-\alpha} \right)$ in front of $x_1^{1-2\alpha}$ in Lemma 4.8 converges to to finite limit as $\alpha \to 0$, while the coefficient $\frac{1}{6 \cdot 20^\alpha \alpha}$ in Lemma 4.9 tends to infinity. On the other hand, when $\alpha \to \frac{1}{2}$, the singularity in the Biot–Savart kernel is strong and getting close to non-integrable. Then it becomes important that the "bad" term integration region contains an angle π range near the singularity, while the "good" region only $\frac{\pi}{4}$. For this reason, controlling the "bad" term for larger values of α is problematic—although there is no reason why there cannot be a different, more clever argument achieving this goal.

Fig. 4.5 The segments I_1 and I_2 and the sets Ω_3 and $K(t_0)$

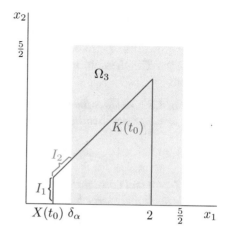

It is straightforward to check that the dominance of the "good" term over "bad" one extends to the range $\alpha \in (0, \frac{1}{24})$, and that in this range we get as a result

$$u_1(x, t) \leq -\frac{1}{50\alpha}x_1^{1-2\alpha} \tag{4.8.15}$$

for $x = (x_1, x_2)$ such that $x_1 \leq \delta_\alpha$ and $x_1 \geq x_2$. A similar bound can be proved showing that

$$u_2(x, t) \geq \frac{1}{50\alpha}x_2^{1-2\alpha} \tag{4.8.16}$$

for $x = (x_1, x_2)$ such that $x_2 \leq \delta_\alpha$ and $x_1 \leq x_2$.

The proof is completed by a contradiction argument, where we assume that the barrier $K(t)$ catches up with the patch $\Omega(t)$ at some time $t = t_0 < T$ of first contact. Taking ϵ sufficiently small compared to δ_α from Lemma 4.9, we can make sure the contact can only happen on the intervals I_1 and I_2 along the boundary of the barrier $K(t_0)$ appearing on Fig. 4.5. But then bounds (4.8.15), (4.8.16) and the evolution of the barrier prescription (4.8.9) lead to the conclusion that the barrier should have been crossed at $t < t_0$, yielding a contradiction.

4.9 The SQG Equation: Smooth Solutions

More details on material presented in this section can be found in [28]. As we already mentioned above, until recently, there have been no examples of smooth solutions to the SQG equation exhibiting infinite in time growth of derivatives. The best known result [35] involved only finite time bursts of growth. In this section, we outline the argument of [28] where examples of SQG solutions with infinite—

exponential in time—growth of derivatives have been obtained. The construction involves a mix of ideas from [36], from the work of Zlatos [64] yielding exponential in time growth for periodic solutions of the 2D Euler equation, and some new ingredients. The theorem below involves the whole range of the modified SQG equations.

Theorem 4.8 *Consider the modified SQG equations* (4.8.1) *in periodic setting. For all* $0 < \alpha < 1$, *there exist initial data* ω_0 *such that*

$$\sup_{t \leq T} \|\nabla^2 \omega(\cdot, t)\|_{L^\infty} \geq \exp(\gamma T), \tag{4.9.1}$$

for all $T > 0$ *and constant* $\gamma > 0$ *that may depend on* ω_0 *and* α. *This constant can be made arbitrarily large by picking* ω_0 *appropriately.*

Note that if $\alpha > 1/2$, the models considered are actually more singular then the SQG equation. Even local well-posedness in this case is not obvious, but has been established in [5]. Also, observe that we do not prove global regularity of the solutions in these examples—solutions that blow up in finite time will also satisfy (4.9.1).

As before, we will make use of symmetries and assume that the initial data (and so the solution) is odd with respect to both x_1 and x_2. We will also need a certain degenerate structure, with solution vanishing to higher order near the x_2 axis. The following lemma shows that this property is preserved in time.

Lemma 4.10 *Suppose that in addition to being odd in* x_1 *and* x_2 *and periodic, the initial data* ω_0 *also satisfies* $\partial_{x_1} \omega_0(0, x_2) = 0$ *for all* x_2. *Then the solution* $\omega(x, t)$, *while it remains smooth, also satisfies* $\partial_{x_1} \omega(0, x_2, t) = 0$.

The lemma is proved by differentiation of the equation and direct analysis. Note that all even derivatives of ω also vanish when $x_1 = 0$ due to odd symmetry.

Next, we need to carefully analyze the Biot–Savart law, which we state below under assumption of oddness in both x_1 and x_2.

$$u_1(x) = \int_0^\infty \int_0^\infty \left(\frac{x_2 - y_2}{|x - y|^{2+2\alpha}} - \frac{x_2 - y_2}{|\tilde{x} - y|^{2+2\alpha}} - \frac{x_2 + y_2}{|\bar{x} - y|^{2+2\alpha}} + \frac{x_2 + y_2}{|x + y|^{2+2\alpha}} \right) \omega(y) \, dy_1 dy_2,$$
$$\tag{4.9.2}$$

$$u_2(x) = -\int_0^\infty \int_0^\infty \left(\frac{x_1 - y_1}{|x - y|^{2+2\alpha}} - \frac{x_1 - y_1}{|\tilde{x} - y|^{2+2\alpha}} - \frac{x_1 + y_1}{|\bar{x} - y|^{2+2\alpha}} + \frac{x_1 + y_1}{|x + y|^{2+2\alpha}} \right) \omega(y) \, dy_1 dy_2.$$
$$\tag{4.9.3}$$

Here $\tilde{x} = (-x_1, x_2)$, $\bar{x} = (x_1, -x_2)$, and the function ω is extended to the entire plane by periodicity. We will later see that the integral converges absolutely at infinity if $\alpha > 0$. Near the singularity $x = y$, the convergence is understood in the principal value sense if $\alpha \geq 1/2$. In what follows, we will denote the kernels in the integrals (4.9.2) and (4.9.3) by $K_1(x, y)$ and $K_2(x, y)$ respectively.

Let $L \geq 1$ be a constant that we will eventually choose to be large enough. The first estimate addresses the contribution of the near field $y_1, y_2 \leq L|x|$ to the Biot–Savart law provided that we have control of $\|\nabla^2 \omega\|_{L^\infty}$. All the inequalities we show in the rest of this section, similarly to Lemma 4.10, assume that the solution remains smooth at times where these inequalities apply.

Lemma 4.11 *Assume that ω is odd with respect to both x_1 and x_2, periodic and smooth. Take $L \geq 2$, and suppose $L|x| \leq 1$. Denote*

$$u_j^{near}(x) = \int_{[0,L|x|]^2} K_j(x, y)\omega(y)\, dy.$$

Then we have

$$|u_j^{near}(x)| \leq Cx_j|x|^{2-2\alpha}L^{2-2\alpha}\|\nabla^2\omega\|_{L^\infty}. \tag{4.9.4}$$

The lemma is proved by about a page of estimates. The control of the second derivative of ω is used, in particular, to estimate the principal value singularity. The key observation is that in the integral

$$\left| P.V. \int_0^{L|x|} dy_1 \int_0^{2x_2} dy_2 \frac{(x_2 - y_2)(|\tilde{x} - y|^{2+2\alpha} - |x - y|^{2+2\alpha})}{\tilde{x} - y|^{2+2\alpha}|x - y|^{2+2\alpha}}\omega(y, t) \right|$$

the kernel is odd with respect to $y_2 = x_2$ line, so $\omega(y, t)$ can be replaced by $\omega(y_1, y_2, t) - \omega(y_1, x_2, t)$. The latter difference can be estimated using second order derivatives by mean value theorem, and the order of the singularity can be reduced to integrable. There are of course more terms to estimate but their control is more straightforward.

The next result records an important property of the Biot–Savart law that makes contribution of the $L|x| \leq |y| \lesssim 1$ region of the central cell to u_1 and u_2 nearly identical when L is large.

Proposition 4.4 *Let L be a parameter and x be such that $L|x| \leq 1$. Assume that ω is odd with respect to both x_1 and x_2, $\omega(x) \geq 0$ in $[0, \pi)^2$, and is positive on a set of measure greater than $(L|x|)^2$. Let us define*

$$u_j^{med}(x) = \int_{[0,\pi)^2 \setminus [0,L|x|]^2} K_j(x, y)\omega(y)\, dy.$$

Then for all sufficiently large $L \geq L_0 \geq 2$ and x such that $L|x| \leq 1$ we have that

$$1 - BL^{-1} \leq -\frac{u_1^{med}(x)x_2}{x_1 u_2^{med}(x)} \leq 1 + BL^{-1}, \tag{4.9.5}$$

with some universal constant B.

The bound (4.9.5) follows from more informative pointwise bound for the Biot–Savart kernel. A direct computation shows that in the region $y_1, y_2 \geq L|x|$, we have

$$K_1(x, y) = -8(1 + \alpha)x_1 y_1 y_2 |y|^{-4-2\alpha}(1 + f_1(x, y)), \tag{4.9.6}$$

and

$$K_2(x, y) = 8(1 + \alpha)x_2 y_1 y_2 |y|^{-4-2\alpha}(1 + f_2(x, y)), \tag{4.9.7}$$

where $|f_{1,2}(x, y)| \leq AL^{-1}$ with some universal constant A.

Lemma 4.11 and Proposition 4.4 control contribution to the Biot–Savart law of the central period cell. It turns out that the contribution of the rest of the cells is regular near the origin.

Lemma 4.12 *Suppose that* $|x| \leq 1$*. Define*

$$u_j^{far}(x) = \int_{[0,\infty)^2 \setminus [0,\pi)^2} K_j(x, y)\omega(y)\,dy.$$

Then

$$|u_j^{far}(x)| \leq C(\alpha)x_j \|\omega\|_{L^\infty}. \tag{4.9.8}$$

The final estimate we need is a lower bound on the absolute value of the velocity components $(-1)^j u_j^{med}$, $j = 1, 2$, near the origin provided certain assumptions on the structure of vorticity.

Lemma 4.13 *There exists a constant* $1 > \delta_0 > 0$ *such that if* $\delta \leq \delta_0$*, the following is true. Suppose, in addition to symmetry assumptions made above, that we have* $1 \geq \omega_0(x) \geq 0$ *on* $[0, \pi)^2$ *and that* $\omega_0(x) = 1$ *if* $\delta \leq x_{1,2} \leq \pi - \delta$*. Then for all* x *and* $L \geq L_0$ *such that* $L|x| \leq \delta$*, we have that*

$$(-1)^j u_j^{med}(x, t) \geq cx_j \delta^{-\alpha}. \tag{4.9.9}$$

The proof of the lemma uses the area preserving property of the flow and the fact that vorticity is conserved along the trajectories, as well as the estimates (4.9.6), (4.9.7).

Now we have all the tools to sketch the proof of Theorem 4.8.

The initial data ω_0 will be chosen as follows. First, as we already discussed, ω_0 is odd with respect to both x_1 and x_2, $1 \geq \omega_0(x) \geq 0$ in $[0, \pi)^2$ and it equals 1 in this region, apart from a region of width $\leq \delta$ along the boundary. The parameter $\delta \leq \delta_0 < 1$ will be fixed later. We also require $\partial_{x_1}\omega_0(0, x_2) = 0$ for all x_2, a condition that is preserved for all times while the solution stays smooth by Lemma 4.10. Finally, we assume that in a small neighborhood of the origin of order

$\sim\delta$ we have $\omega_0(x_1, x_2) = \delta^{-4}x_1^3 x_2$. Note that $\partial_{x_1}^2 \omega_0(0, x_2) = 0$ by oddness, so this is the "maximal" behavior of ω_0 under our degeneracy condition.

Fix arbitrary $T \geq 1$; for small T the result follows automatically as $\|\nabla^2\omega(\cdot, t)\|_{L^\infty} \geq c\delta^{-2}$ for all times. Take $x_1^0 = e^{-T\delta^{-\alpha/2}}$ and $x_2^0 = (x_1^0)^\beta$ where $\beta > 1$ will be chosen later. Observe that

$$\omega_0(x_1^0, x_2^0) = \delta^{-4}(x_1^0)^{3+\beta} = \delta^{-4}e^{-(3+\beta)T\delta^{-\alpha/2}}.$$

Consider the trajectory $(x_1(t), x_2(t))$ originating at (x_1^0, x_2^0). We will track this trajectory until either time reaches T, or $x_2(t)$ reaches x_1^0, or $\|\nabla^2\omega(\cdot, t)\|_{L^\infty}$ becomes large enough to satisfy the lower bound we seek.

Let us denote

$$T_0 = \min\left(T, \min\{t : x_2(t) = x_1^0\}, \min\{t : \|\nabla^2\omega(\cdot, t)\|_{L^\infty} \geq \exp(cT)\}\right).$$

Note that for all $t \leq T_0$, we have $x_2(t) \leq x_1^0$.

The first step is to notice that for all $t \leq T_0$, if the contribution of u^{far} and u^{near} ever becomes of comparable size relatively to u^{med}, and if the parameters δ and L are appropriately chosen, then the growth condition we seek must be satisfied. A specific condition sufficient for growth estimate is

$$|u_j^{near}(x(t), t)| + |u_j^{far}(x(t), t)| \geq L^{-1}(-1)^j u_j^{med}(x(t), t). \tag{4.9.10}$$

This can be derived from Lemmas 4.11, 4.12, and 4.13, and the definition of x_1^0. Hence we can from now on assume that (4.9.10) never holds for $t \leq T_0$; otherwise we are done.

The second step is to observe that if u^{med} does dominate for $t \leq T_0$ and $T_0 = T$, then the exponential growth that we seek would follow simply from Lemma 4.13 as well as conservation of vorticity along trajectories.

This leaves the most interesting case where $T_0 < T$ due to $x_2(t)$ reaching the value x_1^0. The main danger is that u_2 somehow happens to be much more efficient in pushing the trajectory away from the origin than u_1 in compressing it towards the origin. But such scenario is prevented by Proposition 4.4, which basically says that when $u_{1,2}^{med}$ provide dominant contributions to $u_{1,2}$ then these contributions coincide to the main order, and the trajectory is almost precisely a hyperbola. Here is a sketch of the detailed estimate. Using that $u_{1,2}^{med}$ give dominant contributions to $u_{1,2}$ and Proposition 4.4 one can obtain

$$-\frac{u_1(x(t), t)}{x_1(t)} \geq (1 - 2BL^{-1})\frac{u_2(x(t), t)}{x_2(t)} \tag{4.9.11}$$

provided that $B > 2$ and L is sufficiently large that can always be arranged. Therefore

$$\frac{x_1^0}{x_1(T_0)} = e^{-\int_0^{T_0} \frac{u_1(x(t),t)}{x_1(t)} dt} \geq e^{(1-2BL^{-1})\int_0^{T_0} \frac{u_2(x(t),t)}{x_2(t)} dt} =$$

$$\left(\frac{x_2(T_0)}{x_2^0}\right)^{1-2BL^{-1}} = (x_1^0)^{(1-\beta)(1-2BL^{-1})}.$$

Here we used that $x_2^0 = (x_1^0)^\beta$ and $x_2(T_0) = x_1^0$. It follows that

$$x_1(T_0) \leq (x_1^0)^{\beta(1-2BL^{-1})+2BL^{-1}}.$$

This implies that

$$\|\partial_{x_1 x_1}^2 \omega(\cdot, T_0)\|_{L^\infty} \geq 2\omega(x_1(T_0), x_2(T_0), T_0) x_1(T_0)^{-2} \geq \delta^{-4}(x_1^0)^{3+\beta-2\beta(1-2BL^{-1})} \geq$$

$$\delta^{-4}(x_1^0)^{3-\beta+4\beta BL^{-1}} = \delta^{-4} e^{\delta^{-\alpha/2}(\beta-3-4\beta BL^{-1})T}.$$

Now one select β (say $\beta = 5$ would work), L, and δ so that all parts of the argument are valid, obtaining exponential growth we seek in this final step as well.

Acknowledgments The author acknowledges partial support of the NSF-DMS grant 1848790 and 2006372, and thanks the organizers of the 2019 CIME summer school on fluid mechanics for the kind invitation to give lectures and excellent organization.

References

1. H. Bahouri, J.-Y. Chemin, Équations de transport relatives á des champs de vecteurs non-Lipschitziens et mécanique des uides. (French) [Transport equations for non-Lipschitz vector fields and fluid mechanics]. Arch. Rational Mech. Anal. **127**(2), 159–181 (1994)
2. J.T. Beale, T. Kato, A. Majda, Remarks on the breakdown of smooth solutions of the 3D Euler equations. Commun. Math. Phys. **94**, 61–66 (1984)
3. A. Bertozzi, P. Constantin, Global regularity for vortex patches. Commun. Math. Phys. **152**, 19–28 (1993)
4. M.P. Brenner, S. Hormoz, A. Pumir, Potential singularity mechanism for the Euler equations. Phys. Rev. Fluids **1**, 084503 (2016)
5. D. Chae, P. Constantin, D. Cordoba, F. Gancedo, J. Wu, Generalized surface quasi-geostrophic equations with singular velocities. Commun. Pure Appl. Math. **65**(8), 1037–1066 (2012)
6. J.-Y. Chemin, Persistance de structures geometriques dans les fluides incompressibles bidimensionnels. Ann. Ecol. Norm. Super. **26**, 1–26 (1993)
7. J. Chen, T. Hou, Finite time blowup of 2D Boussinesq and 3D Euler equations with $C^{1,\alpha}$ velocity and boundary (2019). Preprint arXiv:1910.00173
8. J. Chen, T. Hou, D. Huang, On the finite time blowup of the De Gregorio model for the 3D Euler equation (2019). Preprint arXiv:1905.06387

9. K. Choi, A. Kiselev, Y. Yao, Finite time blow up for a 1D model of 2D Boussinesq system. Commun. Math. Phys. **334**(3), 1667–1679 (2015)
10. K. Choi, T. Hou, A. Kiselev, G. Luo, V. Sverak, Y. Yao, On the finite-time blowup of a 1D model for the 3D axi-symmetric Euler equations. Commun. Pure Appl. Math. **70**, 2218–2243 (2017)
11. E.A. Coddington, N. Levinson, *Theory of Ordinary Differential Equations* (McGraw-Hill, New York, 1955)
12. P. Constantin, A. Majda, E. Tabak, Formation of strong fronts in the 2D quasi-geostrophic thermal active scalar. Nonlinearity **7**, 1495–1533 (1994)
13. D. Cordoba, Nonexistence of simple hyperbolic blow up for the quasi-geostrophic equation. Ann. Math. **148**, 1135–1152 (1998)
14. D. Cordoba, C. Fefferman, Growth of solutions for QG and 2D Euler equations. J. Am. Math. Soc. **15**, 665–670 (2002)
15. D. Cordoba, M.A. Fontelos, A.M. Mancho, J.L. Rodrigo, Evidence of singularities for a family of contour dynamics equations. Proc. Natl. Acad. Sci. U.S.A. **102**, 5949–5952 (2005)
16. S. Denisov, Infinite superlinear growth of the gradient for the two-dimensional Euler equation. Discrete Contin. Dyn. Syst. A **23**(3), 755–764 (2009)
17. S. Denisov, Double-exponential growth of the vorticity gradient for the two-dimensional Euler equation. Proc. Am. Math. Soc. **143**(3), 1199–1210 (2015)
18. T. Do, A. Kiselev, X. Xu, Stability of blow up for a 1D model of axi-symmetric 3D Euler equation. J. Nonlinear Sci. **28**, 2127–2152 (2018)
19. A. Einstein, The causes of the formation of meanders in the courses of rivers and of the so-called Baer's law, Die Naturwissenschaften 14 (1926); English translation in Ideas and Opinions (1954). http://people.ucalgary.ca/~kmuldrew/river.html
20. T. Elgindi, I.-J. Jeong, Finite-time singularity formation for strong solutions to the Boussinesq system. Preprint arXiv:1708.02724
21. T. Elgindi, I.-J. Jeong, Finite-time Singularity formation for strong solutions to the axi-symmetric 3D Euler equations. Preprint arXiv:1802.09936
22. T. Elgindi, Finite-time singularity formation for $C^{1,\alpha}$ solutions to the incompressible Euler equations on $\mathbb{R}^3$. Preprint arXiv:1904.04795
23. A. Enciso, M.A. García-Ferrero, D. Peralta-Salas, The Biot–Savart operator of a bounded domain. J. Math. Pures Appl. **119**, 85–113 (2018)
24. L. Euler, in *Principes généraux du mouvement des fluides*. Mémoires de L'Académie Royale des Sciences et des Belles-Lettres de Berlin, vol. 11 (1755, 1757), pp. 217–273
25. U. Frisch, *Turbulence. The Legacy of A. N. Kolmogorov* (Cambridge University Press, Cambridge, 1995)
26. F. Gancedo, Existence for the α-patch model and the QG sharp front in Sobolev spaces. Adv. Math. **217**, 2569–2598 (2008)
27. F. Gancedo, N. Patel, On the local existence and blow-up for generalized SQG patches. Preprint arXiv:1811.00530
28. S. He, A. Kiselev, Small scale creation for solutions of the SQG equation. Preprint arXiv:1903.07485
29. I. Held, R. Pierrehumbert, S. Garner, K. Swanson, Surface quasi-geostrophic dynamics. J. Fluid Mech. **282**, 1–20 (1995)
30. V. Hoang, B. Orcan, M. Radosz, H. Yang, Blowup with vorticity control for a 2D model of Boussinesq equations. J. Differ. Equ. **264**, 7328–7356 (2018)
31. E. Hölder, Über die unbeschränkte Fortsetzbarkeit einer stetigen ebenen Bewegung in einer unbegrenzten inkompressiblen Flüssigkeit. Math. Z. **37**, 727–738 (1933)
32. T.Y. Hou, P. Liu, Self-similar singularity of a 1D model for the 3D axisymmetric Euler equations. Res. Math. Sci. **2**, 5 (2015)
33. A. Kiselev, Small scales and singularity formation in fluid dynamics, in *Proceedings of the International Congress of Mathematicians—Rio de Janeiro 2018*. Invited Lectures, vol. III (World Sci. Publ., Hackensack, 2018), pp. 2363–2390

34. A. Kiselev, C. Li, Global regularity and fast small scale formation for Euler patch equation in a smooth domain. Partial Differ. Equ. **44**(4), 279–308 (2019)
35. A. Kiselev, F. Nazarov, in *A Simple Energy Pump for the Periodic 2D Surface Quasi-Geostrophic Equation*. Abel Symphony, vol. 7 (Springer, Heidelberg, 2012), pp. 175–179
36. A. Kiselev, V. Sverak, Small scale creation for solutions of the incompressible two-dimensional Euler equation. Ann. Math. **180**, 1205–1220 (2014)
37. A. Kiselev, L. Ryzhik, Y. Yao, A. Zlatos, Finite time singularity for the modified SQG patch equation. Ann. Math. **184**(3), 909–948 (2016)
38. A. Kiselev, J.-M. Roquejoffre, L. Ryzhik, *Appetizers in Nonlinear PDE*. http://math.stanford.edu/~ryzhik/STANFORD/STANF272-15/notes-272-15.pdf
39. A. Kiselev, Y. Yao, A. Zlatos, Local regularity for the modified SQG patch equation. Commun. Pure Appl. Math. **70**, 1253–1315 (2017)
40. A. Kiselev, C. Tan, Finite time blow up in the hyperbolic Boussinesq system. Adv. Math. **325**, 34–55 (2018)
41. A. Kiselev, H. Yang, Analysis of a singular Boussinesq model. Res. Math. Sci. **6**(1), 13 (2019)
42. G. Luo, T. Hou, Toward the finite-time blowup of the 3D axisymmetric Euler equations: a numerical investigation. Multiscale Model. Simul. **12**(4), 1722–1776 (2014)
43. A. Majda, A. Bertozzi, *Vorticity and Incompressible Flow* (Cambridge University Press, Cambridge, 2002)
44. A. Majda, *Introduction to PDEs and Waves for the Atmoshpere and Ocean*. Courant Lecture Notes in Mathematics (AMS, Providence, 2003)
45. C. Marchioro, M. Pulvirenti, in *Mathematical Theory of Incompressible Nonviscous Fluids*. Applied Mathematical Sciences, vol. 96 (Springer, New York, 1994)
46. Y. Motoori, S. Goto, Generation mechanism of a hierarchy of vortices in a turbulent boundary layer. J. Fluid Mech. **865**, 1085–1109 (2019)
47. C. Muscalu, W. Schlag, *Classical and Multilinear Harmonic Analysis* (Cambridge University Press, Cambridge, 2013)
48. N.S. Nadirashvili, Wandering solutions of the two-dimensional Euler equation (Russian). Funktsional. Anal. i Prilozhen. **25**, 70–71 (1991); translation in Funct. Anal. Appl. **25**, 220–221 (1991, 1992)
49. J. Pedlosky, *Geophysical Fluid Dynamics* (Springer, New York, 1987)
50. L. Prandtl, *Essentials of Fluid Dynamics* (Blackie and Son, London, 1952)
51. J.L. Rodrigo, On the evolution of sharp fronts for the quasi-geostrophic equation. Commun. Pure Appl. Math., **58**, 821–866 (2005)
52. E.-W. Saw, et al., Experimental characterization of extreme events of inertial dissipation in a turbulent swirling flow. Nat. Commun. **7**, 12466 (2016)
53. R.K. Scott, D.G. Dritschel, Numerical simulation of a self-similar cascade of filament instabilities in the surface quasigeostrophic system. Phys. Rev. Lett. **112**, 144505 (2014)
54. R.K. Scott, D.G. Dritschel, Scale-invariant singularity of the surface quasigeostrophic patch. J. Fluid Mech. **863**, R2 (2019)
55. T. Tao, Finite time blowup for an averaged three-dimensional Navier–Stokes equation. J. Am. Math. Soc. **29**, 601–674 (2016)
56. T. Tao, Finite time blowup for Lagrangian modifications of the three-dimensional Euler equation. Ann. PDE **2**(2), 9 (2016)
57. C. Villani, *The Age of the Earth: When the Earth Was Too Young for Darwin* (ICM, 2018). https://www.youtube.com/watch?v=ObSv67R-uyg
58. E. Weinan, K. Khanin, A. Mazel, Y. Sinai, Invariant measures for Burgers equation with stochastic forcing. Ann. Math. **151**(3), 877–960 (2000)
59. W. Wolibner, Un theorème sur l'existence du mouvement plan d'un uide parfait, homogène, incompressible, pendant un temps infiniment long (French). Mat. Z. **37**, 698–726 (1933)
60. X. Xu, Fast growth of the vorticity gradient in symmetric smooth domains for 2D incompressible ideal flow. J. Math. Anal. Appl. **439**(2), 594–607 (2016)
61. V.I. Yudovich, in *The Loss of Smoothness of the Solutions of the Euler Equation with Time (Russian)*. Dinamika Sploshn. Sredy, vol. 16 (Nestacionarnye Problemy Gidordinamiki, 1974), pp. 71–78

62. V.I. Yudovich, On the loss of smoothness of the solutions of the Euler equations and the inherent instability of flows of an ideal fluid. Chaos, **10**, 705–719 (2000)
63. V.I. Yudovich, Eleven great problems of mathematical hydrodynamics. Mosc. Math. J. **3**, 711–737 (2003)
64. A. Zlatos, Exponential growth of the vorticity gradient for the Euler equation on the torus. Adv. Math. **268**, 396–403 (2015)

LECTURE NOTES IN MATHEMATICS

Editors in Chief: J.-M. Morel, B. Teissier;

Editorial Policy

1. Lecture Notes aim to report new developments in all areas of mathematics and their applications – quickly, informally and at a high level. Mathematical texts analysing new developments in modelling and numerical simulation are welcome.

   Manuscripts should be reasonably self-contained and rounded off. Thus they may, and often will, present not only results of the author but also related work by other people. They may be based on specialised lecture courses. Furthermore, the manuscripts should provide sufficient motivation, examples and applications. This clearly distinguishes Lecture Notes from journal articles or technical reports which normally are very concise. Articles intended for a journal but too long to be accepted by most journals, usually do not have this "lecture notes" character. For similar reasons it is unusual for doctoral theses to be accepted for the Lecture Notes series, though habilitation theses may be appropriate.

2. Besides monographs, multi-author manuscripts resulting from SUMMER SCHOOLS or similar INTENSIVE COURSES are welcome, provided their objective was held to present an active mathematical topic to an audience at the beginning or intermediate graduate level (a list of participants should be provided).

   The resulting manuscript should not be just a collection of course notes, but should require advance planning and coordination among the main lecturers. The subject matter should dictate the structure of the book. This structure should be motivated and explained in a scientific introduction, and the notation, references, index and formulation of results should be, if possible, unified by the editors. Each contribution should have an abstract and an introduction referring to the other contributions. In other words, more preparatory work must go into a multi-authored volume than simply assembling a disparate collection of papers, communicated at the event.

3. Manuscripts should be submitted either online at www.editorialmanager.com/lnm to Springer's mathematics editorial in Heidelberg, or electronically to one of the series editors. Authors should be aware that incomplete or insufficiently close-to-final manuscripts almost always result in longer refereeing times and nevertheless unclear referees' recommendations, making further refereeing of a final draft necessary. The strict minimum amount of material that will be considered should include a detailed outline describing the planned contents of each chapter, a bibliography and several sample chapters. Parallel submission of a manuscript to another publisher while under consideration for LNM is not acceptable and can lead to rejection.

4. In general, **monographs** will be sent out to at least 2 external referees for evaluation.

   A final decision to publish can be made only on the basis of the complete manuscript, however a refereeing process leading to a preliminary decision can be based on a pre-final or incomplete manuscript.

   Volume Editors of **multi-author works** are expected to arrange for the refereeing, to the usual scientific standards, of the individual contributions. If the resulting reports can be forwarded to the LNM Editorial Board, this is very helpful. If no reports are forwarded or if other questions remain unclear in respect of homogeneity etc, the series editors may wish to consult external referees for an overall evaluation of the volume.

5. Manuscripts should in general be submitted in English. Final manuscripts should contain at least 100 pages of mathematical text and should always include

   – a table of contents;
   – an informative introduction, with adequate motivation and perhaps some historical remarks: it should be accessible to a reader not intimately familiar with the topic treated;
   – a subject index: as a rule this is genuinely helpful for the reader.
   – For evaluation purposes, manuscripts should be submitted as pdf files.

6. Careful preparation of the manuscripts will help keep production time short besides ensuring satisfactory appearance of the finished book in print and online. After acceptance of the manuscript authors will be asked to prepare the final LaTeX source files (see LaTeX templates online: https://www.springer.com/gb/authors-editors/book-authors-editors/manuscriptpreparation/5636) plus the corresponding pdf- or zipped ps-file. The LaTeX source files are essential for producing the full-text online version of the book, see http://link.springer.com/bookseries/304 for the existing online volumes of LNM). The technical production of a Lecture Notes volume takes approximately 12 weeks. Additional instructions, if necessary, are available on request from lnm@springer.com.

7. Authors receive a total of 30 free copies of their volume and free access to their book on SpringerLink, but no royalties. They are entitled to a discount of 33.3 % on the price of Springer books purchased for their personal use, if ordering directly from Springer.

8. Commitment to publish is made by a *Publishing Agreement*; contributing authors of multiauthor books are requested to sign a *Consent to Publish form*. Springer-Verlag registers the copyright for each volume. Authors are free to reuse material contained in their LNM volumes in later publications: a brief written (or e-mail) request for formal permission is sufficient.

Addresses:
Professor Jean-Michel Morel, CMLA, École Normale Supérieure de Cachan, France
E-mail: moreljeanmichel@gmail.com

Professor Bernard Teissier, Equipe Géométrie et Dynamique,
Institut de Mathématiques de Jussieu – Paris Rive Gauche, Paris, France
E-mail: bernard.teissier@imj-prg.fr

Springer: Ute McCrory, Mathematics, Heidelberg, Germany,
E-mail: lnm@springer.com

Printed in the United States
By Bookmasters